Studies in Systems, Decision and Control

Volume 96

Series editor

Janusz Kacprzyk, Polish Academy of Sciences, Warsaw, Poland
e-mail: kacprzyk@ibspan.waw.pl

About this Series

The series "Studies in Systems, Decision and Control" (SSDC) covers both new developments and advances, as well as the state of the art, in the various areas of broadly perceived systems, decision making and control- quickly, up to date and with a high quality. The intent is to cover the theory, applications, and perspectives on the state of the art and future developments relevant to systems, decision making, control, complex processes and related areas, as embedded in the fields of engineering, computer science, physics, economics, social and life sciences, as well as the paradigms and methodologies behind them. The series contains monographs, textbooks, lecture notes and edited volumes in systems, decision making and control spanning the areas of Cyber-Physical Systems, Autonomous Systems, Sensor Networks, Control Systems, Energy Systems, Automotive Systems, Biological Systems, Vehicular Networking and Connected Vehicles, Aerospace Systems, Automation, Manufacturing, Smart Grids, Nonlinear Systems, Power Systems, Robotics, Social Systems, Economic Systems and other. Of particular value to both the contributors and the readership are the short publication timeframe and the world-wide distribution and exposure which enable both a wide and rapid dissemination of research output.

More information about this series at http://www.springer.com/series/13304

Ramon Garcia-Hernandez
Michel Lopez-Franco · Edgar N. Sanchez
Alma Y. Alanis · Jose A. Ruz-Hernandez

Decentralized Neural Control: Application to Robotics

 Springer

Ramon Garcia-Hernandez
Universidad Autonoma del Carmen
Ciudad del Carmen
Mexico

Alma Y. Alanis
Universidad de Guadalajara
Guadalajara
Mexico

Michel Lopez-Franco
CINVESTAV Unidad Guadalajara
Zapopan
Mexico

Jose A. Ruz-Hernandez
Universidad Autonoma del Carmen
Ciudad del Carmen
Mexico

Edgar N. Sanchez
CINVESTAV Unidad Guadalajara
Zapopan
Mexico

ISSN 2198-4182   ISSN 2198-4190   (electronic)
Studies in Systems, Decision and Control
ISBN 978-3-319-85123-5   ISBN 978-3-319-53312-4   (eBook)
DOI 10.1007/978-3-319-53312-4

Printed on acid-free paper

This Springer imprint is published by Springer Nature
The registered company is Springer International Publishing AG
The registered company address is: Gewerbestrasse 11, 6330 Cham, Switzerland

Preface

The stabilization of large-scale systems has been extensively studied over last years. In the modeling of large-scale systems which are composed of interconnected subsystems, the significant uncertainty is represented by the interconnections among the subsystems. To achieve realizable control of large-scale systems, decentralized controllers must be used. It means that the structure and parameters of decentralized control have to be designed using mathematical models of subsystems in such a way that the achievement of the control goal for the overall system is guaranteed and each local controllers is independent of other large-scale system local controllers of the interconnected system. For nonlinear systems, these requirements demand new approaches to the design of decentralized control. Additional obstacles may be caused by uncertainty which complicates both design of control system and its justification.

Neural networks (NN) have become a well-established methodology as exemplified by their applications to identification and control of general nonlinear and complex systems; the use of high order neural networks for modeling and learning has recently increased. Using neural networks, control algorithms can be developed to be robust to uncertainties and modeling errors. The most used NN structures are feedforward networks and recurrent networks. The latter type offers a better-suited tool to model and control of nonlinear systems. There exist different training algorithms for neural networks, which, however, normally encounter some technical problems such as local minima, slow learning, and high sensitivity to initial conditions, among others. As a viable alternative, new training algorithms have been proposed. There already exist publications about regulation and trajectory tracking using neural networks; however, most of those works were developed for continuous-time systems. On the other hand, while extensive literature is available for linear discrete-time control system, nonlinear discrete-time control design techniques have not been discussed to the same degree. Besides, discrete-time neural networks are better fitted for real-time implementations.

Optimal nonlinear control is related to determining a control law for a given system, such that a cost functional (performance index) is minimized; it is usually formulated as a function of the state and input variables. The major drawback for optimal nonlinear control is the need to solve the associated Hamilton–Jacobi–Bellman (HJB) equation. The HJB equation, as far as we are aware, has not been solved for general nonlinear systems. It has only been solved for the linear regulator problem, for which it is particularly well suited. This book presents a novel inverse optimal control for stabilization and trajectory tracking of discrete-time decentralized nonlinear systems, avoiding the need to solve the associated HJB equation, and minimizing a cost functional. Two approaches are presented; the first one is based on passivity theory and the second one is based on a control Lyapunov function (CLF). It is worth mentioning that if a continuous-time control scheme is real-time implemented, there is no guarantee that it preserves its properties, such as stability margins and adequate performance. Even worse, it is known that continuous-time schemes could become unstable after sampling. There are two advantages to working in a discrete-time framework: (a) appropriate technology can be used to implement digital controllers rather than analog ones; (b) the synthesized controller is directly implemented in a digital processor. Therefore, the control methodology developed for discrete-time nonlinear systems can be implemented in real systems more effectively. In this book, it is considered a class of nonlinear systems, the affine nonlinear systems, which represents a great variety of systems, most of which are approximate discretizations of continuous-time systems. The main characteristic of the inverse optimal control is that the cost functional is determined a posteriori, once the stabilizing feedback control law is established. Important results on inverse optimal control have been proposed for continuous-time linear and nonlinear systems, and the discrete-time inverse optimal control has been analyzed in the frequency domain for linear systems. Different works have illustrated adequate performances of the inverse optimal control due to the fact that this control scheme benefits from adequate stability margins, while the minimization of a cost functional ensures that control effort is not wasted. On the other hand, for realistic situations, a control scheme based on a plant model cannot perform as desired, due to internal and external disturbances, uncertain parameters, and/or unmodeled dynamics. This fact motivates the development of a model based on recurrent high order neural networks (RHONN) in order to identify the dynamics of the plant to be controlled. A RHONN model is easy to implement, has relatively simple structure, and has the capacity to adjust its parameters online. This book establishes a neural inverse optimal controller combining two techniques: (a) inverse optimal control and (b) an online neural identifier, which uses a recurrent neural network, trained with an extended Kalman filter, in order to determine a model for an assumed unknown nonlinear system.

To this end, simulations and real-time implementation for the schemes proposed in this book are presented, validating the theoretical results, using the following prototypes: two DOF robot manipulator, five DOF redundant robot, seven DOF Mitsubishi PA10-7CE robot arm, KUKA youBot mobile robot and Shrimp mobile robot.

Ciudad del Carmen, Campeche, Mexico   Ramon Garcia-Hernandez
September 2016   Michel Lopez-Franco
                                                                            Edgar N. Sanchez
                                                                            Alma Y. Alanis
                                                                     Jose A. Ruz-Hernandez

Acknowledgements

The first and fifth authors thank Universidad Autonoma del Carmen (for its name in Spanish) for encouragement and facilities provided to accomplish this book's publication. Likewise, these authors thank PROMEP (for its name in Spanish, which stands for Program of Improve to Professors) for financial support on project PROMEP/103.5/11/3911.

The authors thank CONACyT (for its name in Spanish, which stands for National Council for Science and Technology), Mexico, for financial support on projects 57801Y, 82084, 86940, CB-106838, CB-131678, CB-156567 and CB-256768.

They also thank CINVESTAV-IPN (for its name in Spanish, which stands for Advanced Studies and Research Center of the National Polytechnic Institute), Mexico, particularly President of the CINVESTAV system, for encouragement and facilities provided to accomplish this book publication. Fourth author thanks CUCEI-UDG (for its name in Spanish, which stands for University Center of Exact Sciences and Engineering of the University of Guadalajara), Mexico and also thanks the support of "Fundacion Marcos Moshinsky".

In addition, they thank Maarouf Saad, École de Technologie Supérieure, Université du Québec, Canada, for allowing to use the five DOF redundant robot and Víctor A. Santibáñez, Miguel A. Llama and Javier Ollervides, Instituto Tecnologico de la Laguna, Torreon, Mexico, for allowing to use the two DOF robot manipulator and carry out the corresponding real-time application.

Ciudad del Carmen, Campeche, Mexico
September 2016

Ramon Garcia-Hernandez
Michel Lopez-Franco
Edgar N. Sanchez
Alma Y. Alanis
Jose A. Ruz-Hernandez

Contents

Acronyms

ANAT   Articulated Nimble Adaptable Trunk
BSFF   Block Strict Feedback Form
CLF   Control Lyapunov Function
DC   Direct Current
DNBC   Decentralized Neural Block Control
DNBS   Decentralized Neural Backstepping Control
DOF   Degrees of Freedom
DSP   Digital Signal Processor
DT   Discrete-Time
EKF   Extended Kalman Filter
HJB   Hamilton–Jacobi–Bellman
HONN   High Order Neural Network
MIMO   Multiple-Input Multiple-Output
MSE   Mean Square Error
NBC   Nonlinear Block Controllable
NN   Neural Network
PWM   Pulse Width Modulation
RHONN   Recurrent High Order Neural Network
RHS   Right-Hand Side
RNN   Recurrent Neural Network
SGUUB   Semiglobally Uniformly Ultimately Bounded
SISO   Single-Input Single-Output

Chapter 1
Introduction

1.1 Preliminaries

Robot manipulators are employed in a wide range of applications such as in manufacturing to move materials, parts, and tools of various types. Future applications will include nonmanufacturing tasks, as in construction, exploration of space, and medical care.

In this context, a variety of control schemes have been proposed in order to guarantee efficient trajectory tracking and stability [36, 38]. Fast advance in computational technology offers new ways for implementing control algorithms within the approach of a centralized control design [15]. However, there is a great challenge to obtain an efficient control for this class of systems, due to its highly nonlinear complex dynamics, the presence of strong interconnections, parameters difficult to determine, and unmodeled dynamics. Considering only the most important terms, the mathematical model obtained requires control algorithms with great number of mathematical operations, which affect the feasibility of real-time implementations.

On the other hand, within the area of control systems theory, for more than three decades, an alternative approach has been developed considering a global system as a set of interconnected subsystems, for which it is possible to design independent controllers, considering only local variables to each subsystem: the so called decentralized control [17, 20]. Decentralized control has been applied in robotics, mainly in cooperative multiple mobile robots and robot manipulators, where it is natural to consider each mobile robot or each part of the manipulator as a subsystem of the whole system. For robot manipulators, each joint and the respective link is considered as a subsystem in order to develop local controllers, which just consider local angular position and angular velocity measurements, and compensate the interconnection effects, usually assumed as disturbances. The resulting controllers are easy to implement for real-time applications [22, 30]. In [12], the authors propose a robust adaptive decentralized control algorithm for trajectory tracking of robot manipulators. The controller, which consists of a PD (Proportional plus Derivative) feedback part and a dynamic compensation part, is designed based on the Lyapunov

© Springer International Publishing Switzerland 2017
R. Garcia-Hernandez et al., *Decentralized Neural Control: Application to Robotics*,
Studies in Systems, Decision and Control 96, DOI 10.1007/978-3-319-53312-4_1

methodology. In [31], a decentralized control of robot manipulators is developed, decoupling the dynamical model of the manipulator in a set of linear subsystems with uncertainties; simulation results for a robot of two joints are included. In [23], an approach of decentralized neural identification and control for robots manipulators is presented using models in discrete-time. In [35], a decentralized control for robot manipulators is reported; it is based on the estimation of each joint dynamics, using feedforward neural networks. A decentralized control scheme, on the basis of a recurrent neural identifier and the block control structure is presented in [37]. This approach was tested only via simulation, with a two degrees of freedom robot manipulator, and with an Articulated Nimble Adaptable Trunk (ANAT) manipulator, with seven degrees of freedom. In [32] the authors present a discrete-time decentralized neural identification and control scheme for large-scale uncertain nonlinear systems, which is developed using recurrent high order neural networks (RHONN); the neural network learning algorithm uses an extended Kalman filter (EKF). The discrete-time control law proposed is based on block control and sliding mode techniques. The control algorithm is first simulated, and then implemented in real time for a two degree of freedom (DOF) planar robot.

In contrast, a mobile robot would be able to travel throughout the manufacturing plant, flexibly applying its talents wherever it is most effective [40]. Fixed manipulators are typically programmed to perform repetitive tasks with perhaps limited use of sensors, whereas mobile robots are less structured in their operation and likely to use more sensors [8]. In applications which are too risky or too demanding for humans, or where a fast response is crucial, multirobot systems can play an important role thanks to their capabilities to cover the whole area. Possible applications are planetary exploration, urban search and rescue, monitoring, surveillance, cleaning, maintenance, among others. In order to successfully perform the tasks, robots require a high degree of autonomy and a good level of cooperation [19, 39].

A mobile robot needs locomotion mechanisms which enable it to move unbounded throughout its environment. However there are a large variety of possible ways to move, and so the selection of a robots approach for locomotion is an important aspect. There are already research robots which can walk, jump, run, slide, skate, swim, fly, and, of course, roll. Most of these locomotion mechanisms have been inspired by their biological counterparts. There is, however, one exception: the actively powered wheel is a human invention which achieves extremely high efficiency on flat ground [40].

An extensive class of controllers have been proposed for mobile robots [1, 9, 11, 21, 26, 34, 42]. Most of these references present only simulation results and the controllers are implemented in continuous time. A common problem when applying standard control theory is that the required parameters are often either unknown at time, or are subject to change during operation. For example, the inertia of a robot as seen at the drive motor has many components, which might include the rotational inertia of the motor rotor, the inertia of gears and shafts, rotational inertia of its tires, the robots empty weight, and its payload; worse yet, there are elements between these components such as bearings, shafts and belts which may have spring constants and friction loads [16].

In the recent literature about adaptive and robust control, numerous approaches have been proposed for the design of nonlinear control systems. Among these, adaptive backstepping constitutes a major design methodology [18, 25]. In [7] an adaptive backstepping neural network control approach is extended to a class of large scale nonlinear output feedback systems, with completely unknown and mismatched interconnections. The idea behind the backstepping approach is that some appropriate functions of state variables are selected recursively as virtual control inputs for lower dimension subsystems of the overall system [29]. Each backstepping stage results in a new virtual control design from the preceding design stages; when the procedure ends, a feedback design for the practical control input results, which achieves the original design objective. The backstepping technique provides a systematic framework for the design of regulation and tracking strategies, suitable for a large class of state feedback linearizable nonlinear systems [6, 25, 27, 28].

The objective of optimal control theory, as applied to nonlinear systems, is to determine the control signals which will force a process to satisfy physical constraints and at the same time to minimize a performance criterion [24]. Unfortunately it is required to solve the associated Hamilton Jacobi Bellman (HJB) equation, which is not an easy task. The target of the inverse optimal control is to avoid the solution of the HJB equation [25]. For the inverse approach, a stabilizing feedback control law is developed and then it is established that this law optimizes a cost functional. The main characteristic of the inverse approach is that the cost function is a posteriori determined from the stabilizing feedback control law [9, 10, 33].

1.2 Motivation

Over the past three decades, the properties of linear interconnected systems have been widely studied [13, 20, 41]. By contrast, the control or modelling of nonlinear interconnected systems have not received the same attention. Due to physical configuration and high dimensionality of interconnected systems, centralized control is neither economically feasible nor even necessary. Decentralized control are free from difficulties due to complexity in design, debugging, data gathering, and storage requirements; they are preferable for interconnected systems than centralized ones [4, 5]. Furthermore, as opposite the centralized approach, the decentralized one has the great advantage to be implemented with parallel processors. These facts motivate the design of discrete-time decentralized neural controllers, using only local information while guaranteeing stability for the whole system.

Therefore, this book deals with the design of decentralized control schemes in order to achieve trajectory tracking for a class of discrete-time nonlinear systems, using neural controllers (neural block control, neural backstepping control and inverse optimal neural control approaches), on the basis of suitable controllers for each subsystem. Accordingly, each subsystem is approximated by an identifier using a discrete-time recurrent high order neural network (RHONN), trained with an extended Kalman filter (EKF) algorithm. The neural identifier scheme is used to provide a decentralized model for the uncertain nonlinear system, and based on this model, then a controller is synthesized in order to achieve trajectory tracking. Applicability of the proposed approaches is illustrated via simulation and real-time implementation for different robotic systems.

1.3   Objectives

The main objectives of this book are stated as follows:

- To synthesize a decentralized scheme for trajectory tracking based on a recurrent high order neural network (RHONN) structure trained with an extended Kalman filter (EKF) to identify the robot dynamics, and to design each controller by the block control and sliding mode technique.
- To synthesize a decentralized scheme for trajectory tracking based on a high order neural network (HONN) structure trained with an EKF, to approximate each controller using the backstepping technique.
- To synthesize a decentralized scheme for stabilization and trajectory tracking based on a recurrent high order neural network (RHONN) structure trained with an extended Kalman filter (EKF) to identify the mobile robots dynamics, and to design each controller using the inverse optimal control technique.
- To develop stability analyses, using the Lyapunov approach, for each one of the proposed schemes.
- To apply the proposed approaches via simulation and real-time implementation for different robotic systems.

1.4   Book Outline

This book is organized as follows:

In *Chap.* 2, mathematical preliminaries are introduced, including stability definitions and the extended Kalman filter training algorithm.

Then *Chap.* 3, presents a discrete-time decentralized control scheme for identification and trajectory tracking. A recurrent high order neural network structure is used to identify the robot manipulator model, and based on this model a discrete-time control law is derived, which combines block control and sliding mode technique. The block control approach is used to design a nonlinear sliding surface such that the resulting sliding mode dynamics is described by a desired linear system [2].

Chapter 4, describes a decentralized neural backstepping approach in order to design a suitable controller for each subsystem. Afterwards, each resulting controller is approximated by a high order neural network [14]. The HONN training is performed on-line by means of an extended Kalman filter [3].

Chapter 5, proposes a decentralized control scheme for stabilization of a nonlinear system using a neural inverse optimal control approach in order to design a suitable controller for each subsystem. Accordingly, each subsystem is approximated by an identifier using a discrete-time recurrent high order neural network (RHONN), trained with an extended Kalman filter (EKF) algorithm.

Chapter 6, discusses a decentralized control scheme for trajectory tracking of mobile robots using a neural inverse optimal control approach in order to design a suitable controller for each subsystem. Accordingly, each subsystem is approximated by an identifier using a discrete-time recurrent high order neural network (RHONN), trained with an extended Kalman filter (EKF) algorithm.

In *Chap.* 7, real-time implementation for the schemes proposed in this book is presented, validating the theoretical results, using a two DOF robot manipulator, a five DOF redundant robot and Shrimp mobile robot. Additionally, simulation results for a seven DOF Mitsubishi PA10-7CE robot arm and KUKA youBot mobile robot are included.

References

1. Akhavan, S., Jamshidi, M.: ANN-based sliding mode control for non-holonomic mobile robots. In: Proceedings of the IEEE International Conference on Control Applications, pp. 664–667. Anchorage, Alaska (2000)
2. Alanis, A.Y., Sanchez, E.N., Loukianov, A.G., Chen, G.: Discrete-time output trajectory tracking by recurrent high-order neural network control. In: Proceedings of the 45th IEEE Conference on Decision and Control, pp. 6367–6372. San Diego, CA, USA (2006)
3. Alanis, A.Y., Sanchez, E.N., Loukianov, A.G.: Discrete-time adaptive backstepping nonlinear control via high-order neural networks. IEEE Trans. Neural Netw. **18**(4), 1185–1195 (2007)
4. Benitez, V.H.: Decentralized Continuous Time Neural Control. Ph.D. thesis, Cinvestav, Unidad Guadalajara, Guadalajara, Jalisco, Mexico (2010)
5. Benitez, V.H., Sanchez, E.N., Loukianov, A.G.: Decentralized adaptive recurrent neural control structure. Eng. Appl. Artif. Intell. **20**(8), 1125–1132 (2007)
6. Campos, J., Lewis, F.L., Selmic, R.: Backlash compensation with filtered prediction in discrete time nonlinear systems using dynamic inversion by neural networks. In: Proceedings of the IEEE International Conference on Robotics and Automation, pp. 1289–1295, San Francisco, CA, USA (2000)
7. Chen, W., Li, J.: Decentralized output-feedback neural control for systems with unknown interconnections. IEEE Trans. Syst. Man Cybern. part B **38**(1), 258–266 (2008)
8. Cook, G.: Mobile Robots: Navigation, Control and Remote Sensing. Wiley, Hoboken (2011)
9. Do, K.D., Jiang, Z.P., Pan, J.: Simultaneous tracking and stabilization of mobile robots: an adaptive approach. IEEE Trans. Autom. Control **49**(7), 1147–1151 (2004)
10. Feldkamp, L.A., Prokhorov, D.V., Feldkamp, T.M.: Simple and conditioned adaptive behavior from Kalman filter trained recurrent networks. Neural Netw. **16**(5), 683–689 (2003)
11. Fierro, R., Lewis, F.L.: Control of a nonholonomic mobile robot using neural networks. IEEE Trans. Neural Netw. **9**(4), 589–600 (1998)
12. Fu, L.-C.: Robust adaptive decentralized control of robot manipulators. IEEE Trans. Autom. Control **37**(1), 106–110 (1992)
13. Gavel, D.T., Siljak, D.D.: Decentralized adaptive control: structural conditions for stability. IEEE Trans. Autom. Control **34**(4), 413–426 (1989)

14. Ge, S.S., Zhang, J., Lee, T.H.: Adaptive neural network control for a class of MIMO nonlinear systems with disturbances in discrete-time. IEEE Trans. Syst. Man Cybern. part B **34**(4), 1630–1645 (2004)
15. Gourdeau, R.: Object-oriented programming for robotic manipulator simulation. IEEE Robot. Autom. **4**(3), 21–29 (1997)
16. Holland, J.: Designing Autonomous Mobile Robots: Inside the Mind of an Intelligent Machine. Newnes, Burlington (2003)
17. Huang, S., Tan, K.K., Lee, T.H.: Decentralized control design for large-scale systems with strong interconnections using neural networks. IEEE Trans. Autom. Control **48**(5), 805–810 (2003)
18. Jagannathan, S.: Control of a class of nonlinear discrete-time systems using multilayer neural networks. IEEE Trans. Neural Netw. **12**(5), 1113–1120 (2001)
19. Jevtic, A., Gutierrez, A., Andina, D., Jamshidi, M.: Distributed bees algorithm for task allocation in swarm of robots. IEEE Syst. J. **6**(2), 296–304 (2012)
20. Jiang, Z.-P.: New results in decentralized adaptive nonlinear control with output feedback. In: Proceedings of the 38th IEEE Conference on Decision and Control, pp. 4772–4777, Phoenix, AZ, USA (1999)
21. Jiang, Z.-P., Nijmeijer, H.: A recursive technique for tracking control of nonholonomic systems in chained form. IEEE Trans. Autom. Control **44**(2), 265–279 (1999)
22. Jin, Y.: Decentralized adaptive fuzzy control of robot manipulators. IEEE Trans. Syst. Man Cybern. Part B **28**(1), 47–57 (1998)
23. Karakasoglu, A., Sudharsanan, S.I., Sundareshan, M.K.: Identification and decentralized adaptive control using dynamical neural networks with application to robotic manipulators. IEEE Trans. Neural Netw. **4**(6), 919–930 (1993)
24. Kirk, D.E.: Optimal Control Theory: An Introduction. Dover Publications, Englewood Cliffs (2004)
25. Krstic, M., Kanellakopoulos, I., Kokotovic, P.V.: Nonlinear and Adaptive Control Design. Wiley, New York (1995)
26. Kumbla, K.K., Jamshidi, M.: Neural network based identification of robot dynamics used for neuro-fuzzy controller. In: Proceedings of the IEEE International Conference on Robotics and Automation, pp. 1118–1123, Albuquerque, NM, USA (1997)
27. Lewis, F.L., Jagannathan, S., Yesildirek, A.: Neural Network Control of Robots Manipulators and Nonlinear Systems. Taylor & Francis Ltd, London (1999)
28. Lewis, F.L., Campos, J., Selmic, R.: Neuro-fuzzy control of industrial systems with actuator nonlinearities. Society for Industrial and Applied Mathematics, Philadelphia, PA, USA (2002)
29. Li, Y., Qiang, S., Zhuang, X., Kaynak, O.: Robust and adaptive backstepping control for nonlinear systems using RBF neural networks. IEEE Trans. Neural Netw. **15**(3), 693–701 (2004)
30. Liu, M.: Decentralized control of robot manipulators: nonlinear and adaptive approaches. IEEE Trans. Autom. Control **44**(2), 357–363 (1999)
31. Ni, M.-L., Er, M.J.: Decentralized control of robot manipulators with coupling and uncertainties. In: Proceedings of the American Control Conference, pp. 3326–3330, Chicago, IL, USA (2000)
32. Ornelas, F., Loukianov, A.G., Sanchez, E.N., Bayro-Corrochano, E.: Decentralized neural identification and control for uncertain nonlinear systems: application to planar robot. J. Frankl. Inst. **347**(6), 1015–1034 (2010)
33. Park, B.S., Yoo, S.J., Park, J.B., Choi, Y.H.: A simple adaptive control approach for trajectory tracking of electrically driven nonholonomic mobile robots. IEEE Trans. Control Syst. Technol. **18**(5), 1199–1206 (2010)
34. Raghavan, V., Jamshidi, M.: Sensor fusion based autonomous mobile robot navigation. In: Proceedings of the IEEE International Conference on System of Systems Engineering, SoSE '07, pp. 1–6, San Antonio, TX, USA (2007)
35. Safaric, R., Rodic, J.: Decentralized neural-network sliding-mode robot controller. In: Proceedings of the 26th Annual Conference on the IEEE Industrial Electronics Society, pp. 906–911, Nagoya, Aichi, Japan (2000)

36. Sanchez, E.N., Ricalde, L.J.: Trajectory tracking via adaptive recurrent neural control with input saturation. In: Proceedings of the International Joint Conference on Neural Networks, pp. 359–364. Portland, OR, USA (2003)
37. Sanchez, E.N., Gaytan, A., Saad, M.: Decentralized neural identification and control for robotics manipulators. In: Proceedings of the IEEE International Symposium on Intelligent Control, pp. 1614–1619, Munich, Germany (2006)
38. Santibañez, V., Kelly, R., Llama, M.A.: A novel global asymptotic stable set-point fuzzy controller with bounded torques for robot manipulators. IEEE Trans. Fuzzy Syst. **13**(3), 362–372 (2005)
39. Shaneyfelt, T., Joordens, M., Nagothu, K., Prevost, J., Kumar, A., Ghazi, S.S.M., Jamshidi, M.: Control and simulation of robotic swarms in heterogeneous environments. In: Proceedings of the IEEE International Conference on Systems, Man and Cybernetics, SMC 2008, pp. 1314–1319, Suntec, Singapore (2008)
40. Siegwart, R., Nourbakhsh, I.R.: Introduction to Autonomous Mobile Robots. Bradford Company, Scituate (2004)
41. Siljak, D.D.: Decentralized Control of Complex Systems. Academic Press, New York (1991)
42. Yang, J.-M., Kim, J.-H.: Sliding mode control for trajectory tracking of nonholonomic wheeled mobile robots. IEEE Trans. Robot. Autom. **15**(3), 578–587 (1999)

Chapter 2
Foundations

2.1 Stability Definitions

Consider a Multiple-Input Multiple-Output (MIMO) nonlinear system:

$$x_{k+1} = F(x_k, u_k) \tag{2.1}$$
$$y_k = h(x_k) \tag{2.2}$$

where $x \in \Re^n$ is the system state, $u \in \Re^m$ is the system input, $y \in \Re^p$ is the system output, and $F \in \Re^n \times \Re^m \rightarrow \Re^n$ is nonlinear function.

Definition 2.1 ([5]) System (2.1) is said to be forced, or to have input. In contrast the system described by an equation without explicit presence of an input u, that is

$$x_{k+1} = F(x_k)$$

is said to be unforced. It can be obtained after selecting the input u as a feedback function of the state

$$u_k = \vartheta(x_k) \tag{2.3}$$

Such substitution eliminates u:

$$x_{k+1} = F(x_k, \vartheta(x_k)) \tag{2.4}$$

and yields an unforced system (2.4) [8].

Definition 2.2 ([5]) The solution of (2.1)–(2.3) is semiglobally uniformly ultimately bounded (SGUUB), if for any Ω, a compact subset of $\Re^n$ and all $x_{k_0} \in \Omega$, there exists an $\varepsilon > 0$ and a number $\mathbf{N}(\varepsilon, x_{k_0})$ such that $\|x_k\| < \varepsilon$ for all $k \geq k_0 + \mathbf{N}$.

In other words, the solution of (2.1) is said to be SGUUB if, for any a priory given (arbitrarily large) bounded set Ω and any a priory given (arbitrarily small)

© Springer International Publishing Switzerland 2017
R. Garcia-Hernandez et al., *Decentralized Neural Control: Application to Robotics*,
Studies in Systems, Decision and Control 96, DOI 10.1007/978-3-319-53312-4_2

set Ω_0, which contains $(0,0)$ as an interior point, there exists a control (2.3) such that every trajectory of the closed loop system starting from Ω enters the set $\Omega_0 = \{x_k | \|x_k\| < \varepsilon\}$, in a finite time and remains in it thereafter [13].

Theorem 2.3 ([5]) *Let $V(x_k)$ be a Lyapunov function for the discrete-time system (2.1), which satisfies the following properties:*

$$\gamma_1(\|x_k\|) \leq V(x_k) \leq \gamma_2(\|x_k\|)$$
$$V(x_{k+1}) - V(x_k) = \Delta V(x_k)$$
$$\leq -\gamma_3(\|x(k)\|) + \gamma_3(\zeta)$$

where ζ is a positive constant, $\gamma_1(\bullet)$ and $\gamma_2(\bullet)$ are strictly increasing functions, and $\gamma_3(\bullet)$ is a continuous, nondecreasing function. Thus if $\Delta V(x_k) < 0$ for $\|x_k\| > \zeta$, then x_k is uniformly ultimately bounded, i.e., there is a time instant k_T, such that $\|x_k\| < \zeta, \forall k < k_T$.

Definition 2.4 ([8]) A subset $S \in \Re^n$ is bounded if there is $r > 0$ such that $\|x\| \leq r$ for all $x \in S$.

Definition 2.5 ([8]) System (2.5) is said to be BIBO stable if for a bounded input u_k, the system produces a bounded output y_k for $0 < k < \infty$

Lemma 2.6 ([17]) *Consider the linear time varying discrete-time system given by*

$$x_{k+1} = A_k x_k + B u_k$$
$$y_k = C x_k \qquad (2.5)$$

where A_k, B and C are appropriately dimensional matrices, $x \in \Re^n$, $u \in \Re^m$ and $y \in \Re^p$.
Let $\Phi(k_1, k_0)$ be the state transition matrix corresponding to A_k for system (2.5), $\Phi(k_1, k_0) = \prod_{k=k_0}^{k=k_1-1} A_k$. If $\Phi(k_1, k_0) < 1 \forall k_1 > k_0 > 0$, then system (2.5) is
(1) globally exponentially stable for the unforced system ($u_k = 0$) and
(2) Bounded Input Bounded Output (BIBO) stable [5, 17].

Theorem 2.7 (Separation Principle) *[12]: The asymptotic stabilization problem of the system (2.1)–(2.2), via estimated state feedback*

$$u_k = \vartheta(\widehat{x}_k)$$
$$\widehat{x}_{k+1} = F(\widehat{x}_k, u_k, y_k) \qquad (2.6)$$

is solvable if and only if the system (2.1)–(2.2) is asymptotically stabilizable and exponentially detectable.

Corollary 2.8 ([12]) *There is an exponential observer for a Lyapunov stable discrete-time nonlinear system (2.1)–(2.2) with $u = 0$ if and only if the linear approximation*

$$x_{k+1} = Ax_k + Bu_k$$

$$y_k = Cx_k \tag{2.7}$$

$$A = \left.\frac{\partial F}{\partial x}\right|_{x=0}, \quad B = \left.\frac{\partial F}{\partial u}\right|_{x=0}, \quad C = \left.\frac{\partial h}{\partial x}\right|_{x=0}$$

of the system (2.1)–(2.2) is detectable.

2.2 Discrete-Time High Order Neural Networks

The use of multilayer neural networks is well known for pattern recognition and for modelling of static systems. The neural network (NN) is trained to learn an input-output map. Theoretical works have proven that, even with one hidden layer, a NN can uniformly approximates any continuous function over a compact domain, provided that the NN has a sufficient number of synaptic connections.

For control tasks, extensions of the first order Hopfield model called Recurrent High Order Neural Networks (RHONN) are used, which presents more interactions among the neurons, as proposed in [14, 16]. Additionally, the RHONN model is very flexible and allows to incorporate to the neural model a priory information about the system structure.

Let consider the problem to identify nonlinear system

$$\chi_{k+1} = F\left(\chi_k, u_k\right) \tag{2.8}$$

where $\chi_k \in \mathfrak{R}^n$ is the state of the system, $u_k \in \mathfrak{R}^m$ is the control input and $F \in \mathfrak{R}^n \times \mathfrak{R}^m \to \mathfrak{R}^n$ is a nonlinear function.

To identify system (2.8), we use the discrete-time RHONN proposed in [18]

$$x_{i,k+1} - w_i^\top \psi_i(x_k, u_k), \quad i = 1, \cdots, n \tag{2.9}$$

where $x_k = [x_{1,k}, x_{2,k} \ldots x_{n,k}]^\top$, x_i is the state of the i-th neuron which identifies the i-th component of state vector χ_k in (2.8), $w_i \in \mathfrak{R}^{L_i}$ is the respective on-line adapted weight vector, and $u_k = [u_{1,k} u_{2,k} \ldots u_{m,k}]^\top$ is the input vector to the neural network; φ_i is an L_i dimensional vector defined as

$$\varphi_i(x_k, u_k) = \begin{bmatrix} \varphi_{i_1} \\ \varphi_{i_2} \\ \vdots \\ \varphi_{i_{L_i}} \end{bmatrix} = \begin{bmatrix} \Pi_{j \in I_1} \xi_{i_j}^{d_{ij}(1)} \\ \Pi_{j \in I_2} \xi_{i_j}^{d_{ij}(2)} \\ \vdots \\ \Pi_{j \in I_{L_i}} \xi_{i_j}^{d_{ij}(L_i)} \end{bmatrix} \tag{2.10}$$

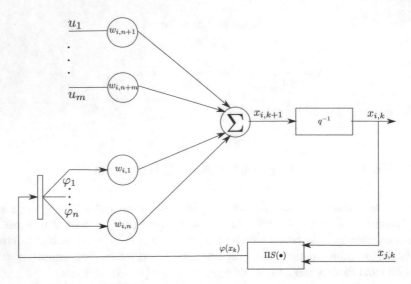

Fig. 2.1 Discrete-Time RHONN

with d_{i_j} nonnegative integers, $j = 1, \ldots, L_i$, L_i is the respective number of high order connections, and $\{I_1, I_2, \cdots, I_{L_i}\}$ a collection of non-ordered subsets of $\{1, 2, \cdots, n + m\}$, n the state dimension, m the number of external inputs, and ξ_i defined as

$$\xi_i = \begin{bmatrix} \xi_{i_1} \\ \vdots \\ \xi_{i_1} \\ \xi_{i_{n+1}} \\ \vdots \\ \xi_{i_{n+m}} \end{bmatrix} = \begin{bmatrix} S(x_{1,k}) \\ \vdots \\ S(x_{n,k}) \\ u_{1,k} \\ \vdots \\ u_{m,k} \end{bmatrix} \tag{2.11}$$

where the sigmoid function $S(\bullet)$ is formulated as

$$S(\varsigma) = \frac{1}{1 + \exp(-\beta\varsigma)}, \quad \beta > 0 \tag{2.12}$$

where β is a constant and ς is any real value variable. Figure 2.1 shows the scheme of a discrete-time RHONN.

Consider the problem to approximate the general discrete-time nonlinear system (2.8), by the following modification of the discrete-time RHONN (2.9) [16]:

$$x_{i,k+1} = w_i^{*\top} \varphi_i(\chi_k) + \varepsilon_{\psi_i} \tag{2.13}$$

where w_i are the adjustable weight matrices, ε_{ψ_i} is a bounded approximation error, which can be reduced by increasing the number of the adjustable weights [16]. Assume that there exists an ideal weights vector w_i^* such that $\|\varepsilon_{\psi_i}\|$ can be minimized on a compact set $\Omega_{\psi_i} \subset \Re^{L_i}$. The ideal weight vector w_i^* is an artificial quantity required for analytical purpose [16]. In general, it is assumed that this vector exists and is constant but unknown. Let us define its estimate as w_i and the respective weight estimation error as

$$\widetilde{w}_{i,k} = w_{i,k} - w_i^* \tag{2.14}$$

Since w_i^* is constant, then $\widetilde{w}_{i,k+1} - \widetilde{w}_{i,k} = w_{i,k+1} - w_{i,k}$. From (2.9) three possible models of RHONN can be derived:

- *Parallel model.* In this configuration, the feedback connections of the NN come from the NN outputs.

$$x_{i,k+1} = w_i^\top \varphi_i(x_k, u_k) \tag{2.15}$$

- *Series-Parallel model.* In this configuration, the feedback connections of the NN are taken from the real plant.

$$x_{i,k+1} = w_i^\top \varphi_i(\chi_k, u_k) \tag{2.16}$$

- *Feedforward model (HONN).* In this configuration, the connections of the NN come from the input signals.

$$x_{i,k+1} = w_i^\top \varphi_i(u_k) \tag{2.17}$$

where x_k is the neural network state vector, χ_k is the plant state vector and u_k is the input vector to the neural network.

2.3 The EKF Training Algorithm

The best well-known training approach for recurrent neural networks (RNN) is the backpropagation through time learning [22]. However, it is a first order gradient descent method and hence its learning speed can be very slow [10]. Recently, Extended Kalman Filter (EKF) based algorithms have been introduced to train neural networks [4, 18], with improved learning convergence [10]. The EKF training of neural networks, both feedforward and recurrent ones, has proven to be reliable and practical for many applications over the past fifteen years [4]. With the EKF based algorithm, learning convergence and robustness are guaranteed as explained in [18].

The training goal is to find the optimal weight values which minimize the prediction error. The EKF-based training algorithm is described for each i-th neuron by [6]:

$$K_{i,k} = P_{i,k} H_{i,k} M_{i,k}$$
$$w_{i,k+1} = w_{i,k} + \eta_i K_{i,k} e_{i,k} \qquad (2.18)$$
$$P_{i,k+1} = P_{i,k} - K_{i,k} H_{i,k}^{\top} P_{i,k} + Q_{i,k}$$

with

$$M_{i,k} = \left[R_{i,k} + H_{i,k}^{\top} P_{i,k} H_{i,k} \right]^{-1} \qquad (2.19)$$
$$e_{i,k} = \chi_{i,k} - x_{i,k} \qquad (2.20)$$
$$i = 1, 2, \cdots, n$$

where $P_i \in \Re^{L_i \times L_i}$ is the prediction error associated covariance matrix, $e_{i,k}$ is the respective identification error, $w_i \in \Re^{L_i}$ is the weight (state) vector, L_i is the total number of neural network weights, $\chi_i \in \Re$ is the i-th plant state component, $x_i \in \Re$ is the i-th neural state component, η_i is a design parameter, $K_i \in \Re^{L_i}$ is the Kalman gain matrix, $Q_i \in \Re^{L_i \times L_i}$ is the state noise associated covariance matrix, $R_i \in \Re$ is the measurement noise associated covariance matrix, $H_i \in \Re^{L_i}$ is a matrix, for which each entry (H_{ij}) is the derivative of one of the neural network output, (x_i), with respect to one neural network weight, (w_{ij}), as follows

$$H_{ij,k} = \left[\frac{\partial x_{i,k}}{\partial w_{ij,k}} \right]^{\top}_{w_{i,k} = w_{i,k+1}} \qquad (2.21)$$

where $i = 1, ..., n$ and $j = 1, ..., L_i$. As an additional parameter, we introduce the rate learning η_i such that $0 \le \eta_i \le 1$. Usually P_i, Q_i and R_i are initialized as diagonal matrices, with entries $P_i(0)$, $Q_i(0)$ and $R_i(0)$, respectively. We set Q_i and R_i fixed. It is important to note that $H_{i,k}$, $K_{i,k}$ and $P_{i,k}$ for the EKF are bounded [21]. Therefore, there exist constants $\bar{H}_i > 0$, $\bar{K}_i > 0$ and $\bar{P}_i > 0$ such that:

$$\|H_{i,k}\| \le \bar{H}_i$$
$$\|K_{i,k}\| \le \bar{K}_i \qquad (2.22)$$
$$\|P_{i,k}\| \le \bar{P}_i$$

It is worth to notice that L_i refers to the number of high order connections and the dimension of the weight vector.

Remark 2.9 The EKF algorithm is used only to train the neural network weights which become the states to be estimated by the EKF.

Remark 2.10 The neural network approximation error vector ϵ_{z_i} is bounded. This is a well-known NN property [3].

2.4 Optimal Control Introduction

This section closely follows [19]. First, we give briefly details about optimal control methodology and their limitations. Let consider the discrete-time affine in the input nonlinear system:

$$\chi_{k+1} = f(\chi_k) + g(\chi_k)u_k, \quad \chi(0) = \chi_0 \qquad (2.23)$$

where $\chi_k \in \Re^n$ is the state of the system, $u_k \in \Re^m$ is the control input, $f(\chi_k) : \Re^n \to \Re^n$ and $g(\chi_k) : \Re^n \to \Re^{n \times m}$ are smooth maps, the subscript $k \in \mathbb{Z}^+ \cup 0 = \{0, 1, 2, \ldots\}$ stands for the value of the functions and/or variables at the time k. We consider that $\bar{\chi}$ is an isolated equilibrium point of $f(\chi) + g(\chi)\bar{u}$ with $\bar{u}$ constant; that is, $f(\bar{\chi}) + g(\bar{\chi})\bar{u} = \bar{\chi}$. Without loss of generality, we consider $\bar{\chi} = 0$ for an $\bar{u}$ constant, $f(0) = 0$ and $rank\{g(\chi_k)\} = m \; \forall \chi_k \neq 0$.

The following meaningful cost function is associated with system (2.23):

$$\mathscr{J}(\chi_k) = \sum_{n=k}^{\infty} (l(\chi_n) + u_n^\top R(\chi_n)u_n) \qquad (2.24)$$

where $\mathscr{J}(\chi_k) : \Re^n \to \Re^+; l(\chi_k) : \Re^n \to \Re^+$ is a positive semidefinite[1] function and $R(\chi_k) : \Re^n \to \Re^{m \times m}$ is a real symmetric positive definite[2] weighting matrix. The meaningful cost functional (2.24) is a performance measure [9]. The entries of $R(\chi_k)$ may be functions of the system state in order to vary the weighting on control efforts according to the state value [9]. Considering the state feedback control approach, we assume that the full state χ_k is available.

Equation (2.24) can be rewritten as

$$\mathscr{J}(\chi_k) = l(\chi_k) + u_k^\top R(\chi_k)u_k + \sum_{n=k+1}^{\infty} l(\chi_n) + u_n^\top R(\chi_n)u_n \qquad (2.25)$$

$$= l(\chi_k) + u_k^\top R(\chi_k)u_k + \mathscr{J}(\chi_{k+1})$$

where we require the boundary condition $\mathscr{J}(0) = 0$ so that $\mathscr{J}(\chi_k)$ becomes a Lyapunov function [1, 20]. The value of $\mathscr{J}(\chi_k)$, if finite, then it is a function of the initial state χ_0. When $\mathscr{J}(\chi_k)$ is at its minimum, which is denoted as $\mathscr{J}^*(\chi_k)$, it is named the optimal value function, and it could be used as a Lyapunov function, i.e., $\mathscr{J}(\chi_k) \triangleq V(\chi_k)$.

From Bellman's optimality principle [2, 11], it is known that, for the infinite horizon optimization case, the value function $V(\chi_k)$ becomes time invariant and satisfies the discrete-time (DT) Bellman equation [1, 2, 15]

---

[1] A function $l(z)$ is positive semidefinite (or nonnegative definite) function if for all vectors z, $l(z) \geq 0$. In other words, there are vectors z for which $l(z) = 0$, and for all others z, $l(z) \geq 0$ [9].

[2] A real symmetric matrix R is positive definite if $z^\top R z > 0$ for all $z \neq 0$ [9].

$$V(\chi_k) = \min_{u_k} \{l(\chi_k) + u_k^\top R(\chi_k)u_k + V(\chi_{k+1})\} \tag{2.26}$$

where $V(\chi_{k+1})$ depends on both χ_k and u_k by means of χ_{k+1} in (2.23). Note that the DT Bellman equation is solved backward in time [1]. In order to establish the conditions that the optimal control law must satisfy, we define the discrete-time Hamiltonian $\mathscr{H}(\chi_k, u_k)$ [7] as

$$\mathscr{H}(\chi_k, u_k) = l(\chi_k) + u_k^\top R(\chi_k)u_k + V(\chi_{k+1}) - V(\chi_k). \tag{2.27}$$

The Hamiltonian is a method to include the constraint (2.23) for the performance index (2.24), and then, solving the optimal control problem by minimizing the Hamiltonian without constraints [11]. A necessary condition that the optimal control law u_k should satisfy is $\frac{\partial \mathscr{H}(\chi_k, u_k)}{\partial u_k} = 0$ [9], which is equivalent to calculate the gradient of (2.26) right-hand side with respect to u_k, then

$$0 = 2R(\chi_k)u_k + \frac{\partial V(\chi_{k+1})}{\partial u_k} \tag{2.28}$$

$$= 2R(\chi_k)u_k + g^\top(\chi_k)\frac{\partial V(\chi_{k+1})}{\partial \chi_{k+1}}$$

Therefore, the optimal control law is formulated as

$$u_k^* = -\frac{1}{2}R^{-1}(\chi_k)g^\top(\chi_k)\frac{\partial V(\chi_{k+1})}{\partial \chi_{k+1}} \tag{2.29}$$

with the boundary condition $V(0) = 0$; u_k^* is used when we want to emphasize that u_k is optimal. Moreover, if $\mathscr{H}(\chi_k, u_k)$ has a quadratic form in u_k and $R(\chi_k) > 0$, then

$$\frac{\partial^2 \mathscr{H}(\chi_k, u_k)}{\partial u_k^2} > 0$$

holds as a sufficient condition such that optimal control law (2.29) (globally [9]) minimizes $\mathscr{H}(\chi_k, u_k)$ and the performance index (2.24) [11].

Substituting (2.29) into (2.26), we obtain the discrete-time Hamilton-Jacobi-Bellman (HJB) equation described by

$$V(\chi_k) = l(\chi_k) + V(\chi_{k+1}) \tag{2.30}$$

$$+ \frac{1}{4}\frac{\partial V^\top(\chi_{k+1})}{\partial \chi_{k+1}}g(\chi_k)R^{-1}(\chi_k)g^\top(\chi_k)\frac{\partial V(\chi_{k+1})}{\partial \chi_{k+1}}$$

which can be rewritten as

$$0 = l(\chi_k) + V(\chi_{k+1}) - V(\chi_k) \tag{2.31}$$
$$+ \frac{1}{4} \frac{\partial V^\top(\chi_{k+1})}{\partial \chi_{k+1}} g(\chi_k) R^{-1}(\chi_k) g^\top(\chi_k) \frac{\partial V(\chi_{k+1})}{\partial \chi_{k+1}}$$

Solving the HJB partial-differential equation (2.31) is not straightforward; this is one of the main disadvantages of discrete-time optimal control for nonlinear systems. To overcome this problem, we propose the inverse optimal control.

Due to the fact that inverse optimal control is based on a Lyapunov function, we establish the following definitions and theorems:

Definition 2.11 A function $V(\chi_k)$ satisfying $V(\chi_k) \to \infty$ as $\|\chi_k\| \to \infty$ is said to be radially unbounded.

Theorem 2.12 *The equilibrium $\chi_k = 0$ of (2.23) is globally asymptotically stable if there is a function $V : \mathfrak{R}^n \to \mathfrak{R}$ such that (I) V is a positive definite function, radially unbounded, and (II) $-\Delta V(\chi_k)$ is a positive definite function, where $\Delta V(\chi_k) = V(\chi_{k+1}) - V(\chi_k)$.*

Theorem 2.13 *Suppose that there exists a positive definite function $V : \mathfrak{R}^n \to \mathfrak{R}$ and constants $c_1, c_2, c_3 > 0$ and $p > 1$ such that*

$$c_1 \|\chi_k\|^p \le V(\chi_k) \le c_2 \|\chi_k\|^p \tag{2.32}$$
$$\Delta V(\chi_k) \le -c_3 \|\chi_k\|^p, \quad \forall k \ge 0, \quad \forall \chi_k \in \mathfrak{R}^n.$$

Then $\chi_k = 0$ is an exponentially stable equilibrium for system (2.23). Clearly, exponential stability implies asymptotic stability. The converse is, however, not true.

Definition 2.14 Let $V(\chi_k)$ be a radially unbounded function, with $V(\chi_k) > 0$, $\forall \chi_k \ne 0$. and $V(0) = 0$. If for any $\chi_k \in \mathfrak{R}^n$, there exist real values u_k such that

$$\Delta V(\chi_k, u_k) < 0 \tag{2.33}$$

where the Lyapunov difference $\Delta V(\chi_k, u_k)$ is defined as $V(\chi_{k+1}) - V(\chi_k) = V(f(\chi_k) + g(\chi_k)u_k) - V(\chi_k)$. Then $V(\bullet)$ is said to be a "discrete-time Control Lyapunov Function" (CLF) for system (2.23).

References

1. Al-Tamimi, A., Lewis, F.L., Abu-Khalaf, M.: Discrete-time nonlinear HJB solution using approximate dynamic programming: convergence proof. IEEE Trans. Syst. Man Cybern. Part B Cybern. **38**(4), 943–949 (2008)
2. Basar, T., Olsder, G.J.: Dynamic Noncooperative Game Theory. Academic Press, New York (1995)
3. Cybenko, G.: Approximation by superpositions of a sigmoidal function. Math. Control Signals Syst. (MCSS) **2**(4), 304–314 (1989)

4. Feldkamp, L.A., Prokhorov, D.V., Feldkamp, T.M.: Simple and conditioned adaptive behavior from Kalman filter trained recurrent networks. Neural Netw. **16**(5), 683–689 (2003)
5. Ge, S.S., Zhang, J., Lee, T.H.: Adaptive neural network control for a class of MIMO nonlinear systems with disturbances in discrete-time. IEEE Trans. Syst. Man Cybern. Part B Cybern. **34**(4), 1630–1645 (2004)
6. Grover, R., Hwang, P.Y.C.: Introduction to Random Signals and Applied Kalman Filtering. Wiley, New York (1992)
7. Haddad, W.M., Chellaboina, V.-S., Fausz, J.L., Abdallah, C.: Identification and control of dynamical systems using neural networks. J. Frankl. Inst. **335**(5), 827–839 (1998)
8. Khalil, H.K.: Nonlinear Systems. Prentice Hall Inc., New Jersey (1996)
9. Kirk, D.E.: Optimal Control Theory: An Introduction. Dover Publications Inc., New Jersey (2004)
10. Leung, C.-S., Chan, L.-W.: Dual extended Kalman filtering in recurrent neural networks. Neural Netw. **16**(2), 223–239 (2003)
11. Lewis, F.L., Syrmos, V.L.: Optimal Control. Wiley, New York (1995)
12. Lin, W., Byrnes, C.I.: Design of discrete-time nonlinear control systems via smooth feedback. IEEE Trans. Autom. Control **39**(11), 2340–2346 (1994)
13. Lin, Z., Saberi, A.: Robust semi-global stabilization of minimum-phase input-output linearizable systems via partial state and output feedback. In: Proceedings of the American Control Conference, pp. 959–963. Baltimore, MD, USA (1994)
14. Narendra, K.S., Parthasarathy, K.: Identification and control of dynamical systems using neural networks. IEEE Trans. Neural Netw. **1**(1), 4–27 (1990)
15. Ohsawa, T., Bloch, A.M., Leok, M.: Discrete Hamilton-Jacobi theory and discrete optimal control. In: Proceedings of the 49th IEEE Conference on Decision and Control, pp. 5438–5443. Atlanta, GA, USA (2010)
16. Rovithakis, G.A., Christodoulou, M.A.: Adaptive Control with Recurrent High-Order Neural Networks. Springer, Berlin (2000)
17. Rugh, W.J.: Linear System Theory. Prentice Hall Inc., New Jersey (1996)
18. Sanchez, E.N., Alanis, A.Y., Loukianov, A.G.: Discrete-Time High Order Neural Control: Trained with Kalman Filtering. Springer, Berlin (2008)
19. Sanchez, E.N., Ornelas-Tellez, F.: Discrete-Time Inverse Optimal Control for Nonlinear Systems. CRC Press, Boca Raton (2013)
20. Sepulchre, R., Jankovic, M., Kokotovic, P.V.: Constructive Nonlinear Control. Springer, London (1997)
21. Song, Y.D., Zhao, S. Liao, X.H., Zhang, R.: Memory-based control of nonlinear dynamic systems part II- applications. In: Proceedings of the 2006 1ST IEEE Conference on Industrial Electronics and Applications, pp. 1–6, Singapore (2006)
22. Williams, R.J., Zipser, D.: A learning algorithm for continually running fully recurrent neural networks. Neural Comput. **1**(2), 270–280 (1989)

Chapter 3
Decentralized Neural Block Control

3.1 Decentralized Systems in Nonlinear Block Controllable Form

Let consider a class of discrete-time nonlinear perturbed and interconnected system which can be presented in the nonlinear block-controllable (NBC) form [11, 12, 16] consisting of r blocks

$$
\begin{aligned}
\chi_{i,k+1}^1 &= f_i^1(\chi_i^1) + B_i^1(\chi_i^1)\chi_i^2 + \Gamma_{i\ell}^1 \\
\chi_{i,k+1}^2 &= f_i^2(\chi_i^1, \chi_i^2) + B_i^2(\chi_i^1, \chi_i^2)\chi_i^3 + \Gamma_{i\ell}^2 \\
&\vdots \\
\chi_{i,k+1}^j &= f_i^j(\chi_i^1, \chi_i^2, \ldots, \chi_i^j) + B_i^j(\chi_i^1, \chi_i^2, \ldots, \chi_i^j)\chi_i^{j+1} + \Gamma_{i\ell}^j \quad (3.1) \\
&\vdots \\
\chi_{i,k+1}^r &= f_i^r(\chi_i) + B_i^r(\chi_i)u_i + \Gamma_{i\ell}^r
\end{aligned}
$$

where $\chi_i \in \Re^{n_i}$, $\chi_i = \begin{bmatrix} \chi_i^{1\top} & \chi_i^{2\top} & \cdots & \chi_i^{r\top} \end{bmatrix}^\top$ and $\chi_i^j \in \Re^{n_{ij}}$, $\chi_i^j = \begin{bmatrix} \chi_{i1}^j & \chi_{i2}^j & \cdots & \chi_{il}^j \end{bmatrix}^\top$, $i = 1, \ldots, N$; $j = 1, \ldots, r-1$; $l = 1, \ldots, n_{ij}$; N is the number of subsystems, $u_i \in \Re^{m_i}$ is the input vector, the rank of $B_i^j = n_{ij}, \sum_{j=1}^r n_{ij} = n_i, \forall \chi_i^j \in D_{\chi_i^j} \subset \Re^{n_{ij}}$. We assume that f_i^j, B_i^j and Γ_i^j are smooth and bounded functions, $f_i^j(0) = 0$ and $B_i^j(0) = 0$. The integers $n_{i1} \le n_{i2} \le \cdots \le n_{ij} \le m_i$ define the different subsystem structures. The interconnection terms are given by

$$
\Gamma_{i\ell}^1 = \sum_{\ell=1, \ \ell \neq i}^N \gamma_{i\ell}^1(\chi_\ell^1)
$$

© Springer International Publishing Switzerland 2017
R. Garcia-Hernandez et al., *Decentralized Neural Control: Application to Robotics*,
Studies in Systems, Decision and Control 96, DOI 10.1007/978-3-319-53312-4_3

$$\Gamma_{i\ell}^2 = \sum_{\ell=1,\ \ell \neq i}^N \gamma_{i\ell}^2(\chi_\ell^1, \chi_\ell^2)$$

$$\vdots \tag{3.2}$$

$$\Gamma_{i\ell}^j = \sum_{\ell=1,\ \ell \neq i}^N \gamma_{i\ell}^j(\chi_\ell^1, \chi_\ell^2, \dots, \chi_\ell^j)$$

$$\vdots$$

$$\Gamma_{i\ell}^r = \sum_{\ell=1,\ \ell \neq i}^N \gamma_{i\ell}^r(\chi_\ell)$$

where χ_ℓ represents the state vector of the ℓ-th subsystem with $1 \leq \ell \leq N$ and $\ell \neq i$. Terms (3.2) reflect the interaction between the i-th subsystem and the other ones.

3.2   Neural Network Identifier

The following decentralized recurrent high order neural network (RHONN) modified model is proposed to identify (3.1):

$$x_{i,k+1}^1 = w_{i,k}^1 S(\chi_{i,k}^1) + w_i^{'1} \chi_{i,k}^2$$
$$x_{i,k+1}^2 = w_{i,k}^2 S(\chi_{i,k}^1, \chi_{i,k}^2) + w_i^{'2} \chi_{i,k}^3$$

$$\vdots$$

$$x_{i,k+1}^j = w_{i,k}^j S(\chi_{i,k}^1, \chi_{i,k}^2, \dots, \chi_{i,k}^j) + w_i^{'j} \chi_{i,k}^{j+1} \tag{3.3}$$

$$\vdots$$

$$x_{i,k+1}^r = w_{i,k}^r S(\chi_{i,k}^1, \dots, \chi_{i,k}^r) + w_i^{'r} u_{i,k}$$

where $x_i^j = \begin{bmatrix} x_i^1 & x_i^2 & \dots & x_i^r \end{bmatrix}^\top$ is the j-th block neuron state with $i = 1, \dots, N$ and $j = 1, \dots, r - 1$, $w_{i,k}^j$ are the adjustable weights, $w_i^{'j}$ are fixed parameters with $rank(w_i^{'j}) = n_{ij}$, $S(\bullet)$ is the activation function, and $u_{i,k}$ represents the input vector.

It is worth to note that, (3.3) constitutes a series-parallel identifier [5, 9] and does not consider explicitly the interconnection terms, whose effects are compensated by the neural network weights update.

Proposition 3.1 *The tracking of a desired trajectory x_{id}^j, defined in terms of the plant state χ_i^j formulated as (3.1) can be established as the following inequality [6]*

$$\left\| x_{id}^{j} - \chi_{i}^{j} \right\| \le \left\| x_{i}^{j} - \chi_{i}^{j} \right\| + \left\| x_{id}^{j} - x_{i}^{j} \right\| \tag{3.4}$$

where $\|\bullet\|$ stands for the Euclidean norm, $i = 1, \ldots, N$, $j = 1, \ldots, r$; $x_{id}^{j} - \chi_{i}^{j}$ is the system output tracking error, $x_{i}^{j} - \chi_{i}^{j}$ is the output identification error, and $x_{id}^{j} - x_{i}^{j}$ is the RHONN output tracking error.

We establish the following requirements for the neural network tracking and control solution:

Requirement 3.1

$$\lim_{t \to \infty} \left\| x_{i}^{j} - \chi_{i}^{j} \right\| \le \zeta_{i}^{j} \tag{3.5}$$

with ζ_{i}^{j} a small positive constant.

Requirement 3.2

$$\lim_{t \to \infty} \| x_{id}^{j} - x_{i}^{j} \| = 0. \tag{3.6}$$

An on-line decentralized neural identifier based on (3.3) ensures (3.5), while (3.6) is guaranteed by a discrete-time controller developed using the block control and sliding mode technique.

It is possible to establish Proposition 3.1 due to separation principle for discrete-time nonlinear systems [10].

3.3 On-Line Learning Law

We use an EKF-based training algorithm described by [4, 7, 8]:

$$
\begin{aligned}
M_{iq,k}^{j} &= [R_{iq,k}^{j} + H_{iq,k}^{j\top} P_{iq,k}^{j} H_{iq,k}^{j}]^{-1} \\
K_{iq,k}^{j} &= P_{iq,k}^{j} H_{iq,k}^{j} M_{iq,k}^{j} \\
w_{iq,k+1}^{j} &= w_{iq,k}^{j} + \eta_{iq}^{j} K_{iq,k}^{j} e_{iq,k}^{j} \\
P_{iq,k+1}^{j} &= P_{iq,k}^{j} - K_{iq,k}^{j} H_{iq,k}^{j\top} P_{iq,k}^{j} + Q_{iq,k}^{j}
\end{aligned}
\tag{3.7}
$$

with

$$e_{iq,k}^{j} = [\chi_{iq,k}^{j} - x_{iq,k}^{j}] \tag{3.8}$$

where $e_{iq,k}^{j}$ is the identification error, $P_{iq,k}^{j}$ is the state estimation prediction error covariance matrix, $w_{iq,k}^{j}$ is the jq-th weight (state) of the i-th subsystem, η_{iq}^{j} is a design parameter such that $0 \le \eta_{iq}^{j} \le 1$, $\chi_{iq,k}^{j}$ is the jq-th plant state, $x_{iq,k}^{j}$ is the jq-th neural network state, q is the number of states, $K_{iq,k}^{j}$ is the Kalman gain matrix, $Q_{iq,k}^{j}$ is the measurement noise covariance matrix, $R_{iq,k}^{j}$ is the state noise covariance matrix, and $H_{iq,k}^j$ is a matrix, in which each entry of (H_q^j) is the derivative of jq-th neural network state $(x_{iq,k}^j)$, with respect to all adjustable weights (w_{iq}^j), as follows

$$H_{q,k}^j = \left[\frac{\partial x_{iq,k}^j}{\partial w_{iq,k}^j} \right]_{w_{iq,k}^j = w_{iq,k+1}^j}^{\top}, \tag{3.9}$$

where $i = 1, \ldots, N$, $j = 1, \ldots, r_i$ and $q = 1, \ldots, n_{ij}$. Usually P_{iq}^j and Q_{iq}^j are initialized as diagonal matrices, with entries $P_{iq}^j(0)$ and $Q_{iq}^j(0)$, respectively [8]. It is important to remark that $H_{iq,k}^j$, $K_{iq,k}^j$, and $P_{iq,k}^j$ for the EKF are bounded [14].

Then the dynamics of (3.8) can be expressed as

$$e_{iq,k+1}^j = \tilde{w}_{iq,k}^j z_{iq}^j(x_k, u_k) + \varepsilon_{z_{iq}^j} \tag{3.10}$$

on the other hand, the dynamics of weight estimation error $\tilde{w}_{iq,k}^j$ is

$$\tilde{w}_{iq,k+1}^j = \tilde{w}_{iq,k}^j - \eta_{iq}^j K_{iq,k}^j e_{iq,k}^j. \tag{3.11}$$

For the case when i is fixed, the stability analysis for the i-th subsystem of RHONN (3.3) to identify the i-th subsystem of nonlinear plant (3.1), is based on the following theorem.

Theorem 3.2 *The i-th subsystem of RHONN (3.3) trained with the EKF-based algorithm (3.7) to identify the i-subsystem of nonlinear plant (3.1) in absence of interconnections, ensures that the identification error (3.8) is semiglobally uniformly ultimately bounded (SGUUB); moreover, the RHONN weights remain bounded.*

Proof Let consider the Lyapunov function candidate

$$V_{iq,k}^j = \tilde{w}_{iq,k}^{j\top} P_{iq,k}^j \tilde{w}_{iq,k}^j + e_{iq,k}^{j2}. \tag{3.12}$$

$$\begin{aligned}
\Delta V_{iq,k}^j &= V_{iq,k+1}^j - V_{iq,k}^j \\
&= \tilde{w}_{iq,k+1}^{j\top} P_{iq,k+1}^j \tilde{w}_{iq,k+1}^j + e_{iq,k+1}^{j2} \\
&\quad - \tilde{w}_{iq,k}^{j\top} P_{iq,k}^j \tilde{w}_{iq,k}^j + e_{iq,k}^{j2}.
\end{aligned}$$

Using (3.10) and (3.11) in (3.12),

$$\begin{aligned}
\Delta V_{iq,k}^j &= [\tilde{w}_{iq,k}^j - \eta_{iq}^j K_{iq,k}^j e_{iq,k}^j]^{\top} \\
&\quad \times [P_{iq,k}^j - A_{iq,k}^j][\tilde{w}_{iq,k}^j - \eta_{iq}^j K_{iq,k}^j e_{iq,k}^j] \\
&\quad + [\tilde{w}_{iq,k}^j z_{iq}^j(x_k, u_k) + \varepsilon_{z_{iq}^j}]^2 \\
&\quad - \tilde{w}_{iq,k}^{j\top} P_{iq,k}^j \tilde{w}_{iq,k}^j - e_{iq,k}^{j2}.
\end{aligned} \tag{3.13}$$

with $A_{iq,k}^{j} = K_{iq,k}^{j} H_{iq,k}^{j\top} P_{iq,k}^{j} + Q_{iq,k}^{j}$; then (3.13) can be expressed as

$$
\begin{aligned}
\Delta V_{iq,k}^{j} = {} & \tilde{w}_{iq,k}^{j\top} P_{iq,k}^{j} \tilde{w}_{iq,k}^{j} - \eta_{iq}^{j} e_{iq,k}^{j} K_{iq,k}^{j} P_{iq,k}^{j} \tilde{w}_{iq,k}^{j} \\
& - \tilde{w}_{iq,k}^{j\top} A_{iq,k}^{j} \tilde{w}_{iq,k}^{j} + \eta_{iq}^{j} e_{iq,k}^{j} K_{iq,k}^{j\top} A_{iq,k}^{j} \tilde{w}_{iq,k}^{j} \\
& - \eta_{iq}^{j} e_{iq,k}^{j} \tilde{w}_{iq,k}^{j\top} P_{iq,k}^{j} K_{iq,k}^{j} \\
& + \eta_{iq}^{j^2} e_{iq,k}^{j^2} K_{iq,k}^{j\top} P_{iq,k}^{j} K_{iq,k}^{j} \\
& + \eta_{iq}^{j} e_{iq,k}^{j} \tilde{w}_{iq,k}^{j\top} A_{iq,k}^{j} K_{iq,k}^{j} \\
& - \eta_{iq}^{j^2} e_{iq,k}^{j^2} K_{iq,k}^{j\top} A_{iq,k}^{j} K_{iq,k}^{j} \\
& + (\tilde{w}_{iq,k}^{j} z_{iq}^{j}(x_k, u_k))^2 \\
& + 2\varepsilon_{z_{iq}^{j}} \tilde{w}_{iq,k}^{j} z_{iq}^{j}(x_k, u_k) \\
& + \varepsilon_{z_{iq}^{j}}^{2} - \tilde{w}_{iq,k}^{j\top} P_{iq,k}^{j} \tilde{w}_{iq,k}^{j} - e_{iq,k}^{j^2}.
\end{aligned}
\tag{3.14}
$$

Using the inequalities

$$
\begin{aligned}
& X^\top X + Y^\top Y \geq 2X^\top Y \\
& X^\top X + Y^\top Y \geq -2X^\top Y \\
& -\lambda_{\min}(P)X^2 \geq -X^\top P X \geq -\lambda_{\max}(P)X^2
\end{aligned}
\tag{3.15}
$$

which are valid $\forall X, Y \in \Re^n, \forall P \in \Re^{n\times n}, P = P^\top > 0$, (3.14) can be rewritten as

$$
\begin{aligned}
\Delta V_{iq,k}^{j} = {} & - \tilde{w}_{iq,k}^{j\top} A_{iq,k}^{j} \tilde{w}_{iq,k}^{j} - \eta_{iq}^{j} e_{iq,k}^{j^2} K_{iq,k}^{j\top} A_{iq,k}^{j} K_{iq,k}^{j} \\
& + \tilde{w}_{iq,k}^{j\top} \tilde{w}_{iq,k}^{j} + e_{iq,k}^{j^2} \\
& + \eta_{iq}^{j^2} e_{iq,k}^{j^2} K_{iq,k}^{j\top} P_{iq,k}^{j} P_{iq,k}^{j\top} K_{iq,k}^{j} \\
& + \eta_{iq}^{j^2} \tilde{w}_{iq,k}^{j\top} A_{iq,k}^{j} K_{iq,k}^{j\top} A_{iq,k}^{j\top} \tilde{w}_{iq,k}^{j} \\
& + \eta_{iq}^{j^2} e_{iq,k}^{j^2} K_{iq,k}^{j\top} P_{iq,k}^{j} K_{iq,k}^{j} \\
& + 2(\tilde{w}_{iq,k}^{j} z_{iq}^{j}(x_k, u_k))^2 + 2\varepsilon_{z_{iq}^{j}} - e_{iq,k}^{j^2}
\end{aligned}
\tag{3.16}
$$

Then

$$
\begin{aligned}
\Delta V_{iq,k}^{j} \leq {} & - \|\tilde{w}_{iq,k}^{j}\|^2 \lambda_{\min}(\Lambda_{iq,k}^{j}) \\
& - \eta_{iq}^{j^2} e_{iq,k}^{j^2} \|K_{iq,k}^{j}\|^2 \lambda_{\min}(A_{iq,k}^{j}) + \|\tilde{w}_{iq,k}^{j}\|^2 \\
& + \eta_{iq}^{j^2} e_{iq,k}^{j^2} \|K_{iq,k}^{j}\|^2 \lambda_{\max}^{2}(P_{iq,k}^{j}) \\
& + \eta_{iq}^{j^2} \|\tilde{w}_{iq,k}^{j}\|^2 \lambda_{\max}^{2}(A_{iq,k}^{j}) \|K_{iq,k}^{j}\|^2 \\
& + \eta_{iq}^{j^2} e_{iq,k}^{j^2} \|K_{iq,k}^{j}\|^2 \lambda_{\max}(P_{iq,k}^{j}) \\
& + 2\|\tilde{w}_{iq,k}^{j}\|^2 \|z_{iq}^{j}(x_k, u_k)\|^2 + 2\varepsilon_{z_{iq}^{j}}^{2}
\end{aligned}
\tag{3.17}
$$

Defining

$$E_{iq,k}^j = \lambda_{\min}(A_{iq,k}^j) - \eta_{iq}^{j^2}\lambda_{\max}^2(A_{iq,k}^j)\|K_{iq,k}^j\|^2$$
$$- 2\|z_{iq}^j(x_k, u_k)\|^2 - 1$$
$$F_{iq,k}^j = \eta_{iq}^{j^2}\|K_{iq,k}^j\|^2\lambda_{\min}(A_{iq,k}^j) \qquad (3.18)$$
$$- \eta_{iq}^{j^2}\|K_{iq,k}^j\|^2\lambda_{\max}^2(P_{iq,k}^j)$$
$$- \eta_{iq}^{j^2}\|K_{iq,k}^j\|^2\lambda_{\max}(P_{iq,k}^j)$$

and selecting η_{iq}^j, Q_{iq}^j, and R_{iq}^j, such that $E_{iq,k}^j > 0$ and $F_{iq,k}^j > 0$, $\forall k$, (3.17) can be expressed as

$$\Delta V_{iq,k}^j \leq -\|\tilde{w}_{iq,k}^j\|^2 E_{iq,k}^j - |e_{iq,k}^j|^2 F_{iq,k}^j + 2\varepsilon_{z_{iq}^j}^2.$$

Hence, $\Delta V_{iq,k}^j < 0$ when

$$\|\tilde{w}_{iq,k}^j\| > \frac{\sqrt{2}|\varepsilon_{z_{iq}^j}|}{\sqrt{E_{iq,k}^j}} \equiv \kappa_{i1}^j$$

and

$$|e_{iq,k}^j| > \frac{\sqrt{2}|\varepsilon_{z_{iq}^j}|}{\sqrt{F_{iq,k}^j}} \equiv \kappa_{i2}^j$$

Therefore, the weight estimation error $\tilde{w}_{iq,k}^j$ and the identification error $e_{iq,k}^j$ are ultimately bounded by

$$\|\tilde{w}_{iq,k}^j\| \leq b_1 \quad \forall k > k_1$$

and

$$|e_{iq,k}^j| \leq b_2 \quad \forall k > k_2$$

where

$$b_1 = \sqrt{\frac{\lambda_{\max}(P_{iq,k}^j)}{\lambda_{\min}(P_{iq,k}^j)}}\kappa_{i1}^j$$

and

$$b_2 = \kappa_{i2}^j$$

Then, according to Theorem 2.3, the solution of (3.10) and (3.11) are SGUUB.

Now, let consider the RHONN (3.3) which identify the nonlinear plant (3.1) in presence of interconnections.

Theorem 3.3 *Assume that the solution of i-th subsystem of RHONN (3.3) is satisfied by the bounds b_1 and b_2, respectively of Theorem 3.2, then the RHONN (3.3) with $i = 1, \ldots, N$ trained with the EKF-based algorithm (3.7) to identify the nonlinear*

plant (3.1) in presence of interconnections, ensures that the identification error (3.8)
and the RHONN weights are semiglobally uniformly ultimately bounded (SGUUB).

Proof Let $V_k = \sum_{i=1}^{N} \sum_{j=1}^{r_i} \sum_{q=1}^{n_{ij}} V_{iq,k}^j$, then

$$
\begin{aligned}
\Delta V_k &= \sum_{i=1}^{N} \sum_{j=1}^{r_i} \sum_{q=1}^{n_{ij}} \left(\tilde{w}_{iq,k+1}^{j\top} P_{iq,k+1}^j \tilde{w}_{iq,k+1}^j + e_{iq,k+1}^{j^2} \right. \\
&\qquad \left. - \tilde{w}_{iq,k}^{j\top} P_{iq,k}^j \tilde{w}_{iq,k}^j + e_{iq,k}^{j^2} \right) \\
&= \sum_{i=1}^{N} \sum_{j=1}^{r_i} \sum_{q=1}^{n_{ij}} \left([\tilde{w}_{iq,k}^j - \eta_{iq}^j K_{iq,k}^j e_{iq,k}^j]^\top \times [P_{iq,k}^j - A_{iq,k}^j] \right. \\
&\qquad [\tilde{w}_{iq,k}^j - \eta_{iq}^j K_{iq,k}^j e_{iq,k}^j] + [\tilde{w}_{iq,k}^j z_{iq}^j(x_k, u_k) + \varepsilon_{z_{iq}^j}]^2 \\
&\qquad \left. - \tilde{w}_{iq,k}^{j\top} P_{iq,k}^j \tilde{w}_{iq,k}^j - e_{iq,k}^{j^2} \right)
\end{aligned}
\tag{3.19}
$$

Equation (3.19) can be expressed as

$$
\begin{aligned}
\Delta V_k &= \sum_{i=1}^{N} \sum_{j=1}^{r_i} \sum_{q=1}^{n_{ij}} \left(- \tilde{w}_{iq,k}^{j\top} A_{iq,k}^j \tilde{w}_{iq,k}^j - \eta_{iq}^j e_{iq,k}^{j^2} K_{iq,k}^{j\top} A_{iq,k}^j K_{iq,k}^j \right. \\
&\qquad + \tilde{w}_{iq,k}^{j\top} \tilde{w}_{iq,k}^j + e_{iq,k}^{j^2} + \eta_{iq}^{j^2} e_{iq,k}^{j^2} K_{iq,k}^{j\top} P_{iq,k}^j P_{iq,k}^{j\top} K_{iq,k}^j \\
&\qquad + \eta_{iq}^{j^2} \tilde{w}_{iq,k}^{j\top} A_{iq,k}^j K_{iq,k}^{j\top} A_{iq,k}^{j\top} \tilde{w}_{iq,k}^j + \eta_{iq}^{j^2} e_{iq,k}^{j^2} K_{iq,k}^{j\top} P_{iq,k}^j K_{iq,k}^j \\
&\qquad \left. + 2(\tilde{w}_{iq,k}^j z_{iq}^j(x_k, u_k))^2 + 2\varepsilon_{z_{iq}^j} - e_{iq,k}^{j^2} \right)
\end{aligned}
\tag{3.20}
$$

$$
\begin{aligned}
\Delta V_k &\leq \sum_{i=1}^{N} \sum_{j=1}^{r_i} \sum_{q=1}^{n_{ij}} \left(- \|\tilde{w}_{iq,k}^j\|^2 \lambda_{\min}(A_{iq,k}^j) + \eta_{iq}^{j^2} e_{iq,k}^{j^2} \|K_{iq,k}^j\|^2 \lambda_{\max}^2(P_{iq,k}^j) \right. \\
&\qquad - \eta_{iq}^{j^2} e_{iq,k}^{j^2} \|K_{iq,k}^j\|^2 \lambda_{\min}(A_{iq,k}^j) + \|\tilde{w}_{iq,k}^j\|^2 \\
&\qquad + \eta_{iq}^{j^2} \|\tilde{w}_{iq,k}^j\|^2 \lambda_{\max}^2(A_{iq,k}^j) \|K_{iq,k}^j\|^2 + \eta_{iq}^{j^2} e_{iq,k}^{j^2} \|K_{iq,k}^j\|^2 \lambda_{\max}(P_{iq,k}^j) \\
&\qquad \left. + 2\|\tilde{w}_{iq,k}^j\|^2 \|z_{iq}^j(x_k, u_k)\|^2 + 2\varepsilon_{z_{iq}^j}^2 \right)
\end{aligned}
$$

Defining $E_{iq,k}^j$ and $F_{iq,k}^j$ as in (3.18) and selecting η_{iq}^j, Q_{iq}^j, and R_{iq}^j, such that
$E_{iq,k}^j > 0$ and $F_{iq,k}^j > 0$, $\forall k$, (3.20) can be expressed as

$$
\Delta V_{iq,k}^j \leq \sum_{i=1}^{N} \sum_{j=1}^{r_i} \sum_{q=1}^{n_{ij}} \left(- \|\tilde{w}_{iq,k}^j\|^2 E_{iq,k}^j - |e_{iq,k}^j|^2 F_{iq,k}^j + 2\varepsilon_{z_{iq}^j}^2 \right).
$$

Hence, $\Delta V_{iq,k}^{j} < 0$ when

$$\|\tilde{w}_{iq,k}^{j}\| > \frac{\sqrt{2}|\varepsilon_{z_{iq}^{j}}|}{\sqrt{E_{iq,k}^{j}}} \equiv \kappa_{i1}^{j}$$

and

$$|e_{iq,k}^{j}| > \frac{\sqrt{2}|\varepsilon_{z_{iq}^{j}}|}{\sqrt{F_{iq,k}^{j}}} \equiv \kappa_{i2}^{j}$$

Therefore, the weight estimation error $\tilde{w}_{iq,k}^{j}$ and the identification error $e_{iq,k}^{j}$ are ultimately bounded by

$$\|\tilde{w}_{iq,k}^{j}\| \leq b_1 \quad \forall k > k_1$$

and

$$|e_{iq,k}^{j}| \leq b_2 \quad \forall k > k_2$$

where

$$b_1 = \sqrt{\frac{\lambda_{\max}(P_{iq,k}^{j})}{\lambda_{\min}(P_{iq,k}^{j})}} \kappa_{i1}^{j} \quad \text{and} \quad b_2 = \kappa_{i2}^{j}$$

$\forall i = 1, \ldots, N; \ j = 1, \ldots, r_i; \ q = 1, \ldots, n_{ij}.$

3.4 Controller Design

It is worth to mention that the identification error bound (3.5) can be achieve arbitrarily small by means introducing high order terms in the identifier (3.3). Therefore, assume that $\zeta_i = 0$, namely the estimated $x_{i,k}$, is equal to $\chi_{i,k}$ that is

$$x_{i,k} = \chi_{i,k}. \tag{3.21}$$

The case when $\zeta_i \neq 0$ is considered in [1, 3].

The objective is develop a tracking control law for system (3.1). We use a combination of discrete-time neural block control [2, 13] and sliding mode technique [17]. Using the series-parallel model (3.3), we begin defining the tracking error as

$$\mathbf{z}_{i,k}^{1} = x_{i,k}^{1} - x_{i\mathrm{d},k}^{1} \tag{3.22}$$

where $x_{i,k}^{1}$ is the i-th neural network state and $x_{i\mathrm{d},k}^{j}$ is the i-th desired trajectory signal with $i = 1, \ldots, N$ subsystems.

Once the first new variable (3.22) is defined, one step ahead is taken

$$z^1_{i,k+1} = w^1_{i,k}S(\chi^1_{i,k}) + w'^1_i \, \chi^2_{i,k} - x^1_{id,k+1}. \tag{3.23}$$

Equation (3.23) is viewed as a block with state $z^1_{i,k}$ and the state $\chi^2_{i,k}$ is considered as a pseudo-control input, where desired dynamics can be imposed. This equation can be solved with the anticipation of the desired dynamics for this block as

$$\begin{aligned} z^1_{i,k+1} &= w^1_{i,k}S(\chi^1_{i,k}) + w'^1_i \, \chi^2_{i,k} - x^1_{id,k+1} \\ &= k^1_i z^1_{i,k} \end{aligned} \tag{3.24}$$

where $|k^1_i| < 1$; in order to ensure stability of (3.24). Under condition (3.21) $x^2_{id,k}$ is calculated as

$$\begin{aligned} x^2_{id,k} = \chi^2_{i,k} = \tfrac{1}{w'^1_i}\left(-w^1_{i,k}S(\chi^1_{i,k}) + x^1_{id,k+1}\right. \\ \left. +k^1_i z^1_{i,k}\right). \end{aligned} \tag{3.25}$$

Note that the calculated value for state $x^2_{id,k}$ in (3.25) is not its real value; instead of it, represents the desired behavior for $\chi^2_{i,k}$. So, to avoid confusions this desired value of $\chi^2_{i,k}$ is referred as $x^2_{id,k}$ in (3.25).

Proceeding in the same way as for the first block, a second variable in the new coordinates is defined by:

$$z^2_{i,k} = x^2_{i,k} - x^2_{id,k}.$$

Taking one step ahead for $z^2_{i,k}$ yields

$$z^2_{i,k+1} = x^2_{i,k+1} - x^2_{id,k+1}$$

The desired dynamics for this block is imposed as

$$\begin{aligned} z^2_{i,k+1} &= w^2_{i,k}S(\chi^1_{i,k}, \chi^2_{i,k}) + w'^2_i \, \chi^3_{i,k} \\ &\quad -x^2_{id,k+1} \\ &= k^2_i z^2_{i,k} \end{aligned} \tag{3.26}$$

where $|k^2_i| < 1$.

These steps are taken iteratively. At the last step, the known desired variable is $x^r_{id,k}$, and the last new variable is defined as

$$z^r_{i,k} = x^r_{i,k} - x^r_{id,k}.$$

As usually, taking one step ahead yields

$$\mathbf{z}_{i,k+1}^{r} = w_{i,k}^{r} S(\chi_{i,k}^{1}, \dots, \chi_{i,k}^{r}) \\ + w_{i}^{'r} u_{i,k} - x_{id,k+1}^{r}. \tag{3.27}$$

System (3.3) can be represented in the new variables $\mathbf{z}_{i} = \left[\mathbf{z}_{i}^{1\top} \mathbf{z}_{i}^{2\top} \cdots \mathbf{z}_{i}^{r\top} \right]$ as

$$\begin{aligned} \mathbf{z}_{i,k+1}^{1} &= k_{i}^{1} \mathbf{z}_{i,k}^{1} + w_{i}^{'1} \mathbf{z}_{i,k}^{2} \\ \mathbf{z}_{i,k+1}^{2} &= k_{i}^{2} \mathbf{z}_{i,k}^{2} + w_{i}^{'2} \mathbf{z}_{i,k}^{3} \\ &\vdots \\ \mathbf{z}_{i,k+1}^{r-1} &= k_{i}^{r-1} \mathbf{z}_{i,k}^{r-1} + w_{i}^{'(r-1)} \mathbf{z}_{i,k}^{r} \\ \mathbf{z}_{i,k+1}^{r} &= w_{i,k}^{r} S(\chi_{i}^{1}(k), \dots, \chi_{i,k}^{r}) + w_{i}^{'r} u_{i,k} \\ &\quad - x_{id,k+1}^{r}. \end{aligned} \tag{3.28}$$

For a sliding mode control implementation [15], when the control resources are bounded by u_{0i}

$$\left| u_{i,k} \right| \leq u_{0i} \tag{3.29}$$

a sliding manifold and a control law which drives the states toward such manifold must be designed. The sliding manifold is selected as $S_{D_{i,k}} = \mathbf{z}_{i,k}^{r} = 0$; then, system (3.27) is rewritten as follows:

$$S_{D_{i,k+1}} = w_{i,k}^{r} S(\chi_{i,k}^{1}, \dots, \chi_{i,k}^{r}) + w_{i}^{'r} u_{i,k} \\ - x_{id,k+1}^{r}. \tag{3.30}$$

Once the sliding manifold is defined, the next step is to find a control law which takes into consideration the bound (3.29), therefore, the control $u_{i,k}$ is selected as [17]:

$$u_{i,k} = \begin{cases} u_{eq_{i,k}} & \text{for } \left\| u_{eq_{i,k}} \right\| \leq u_{0i} \\ u_{0i} \dfrac{u_{eq_{i,k}}}{\left\| u_{eq_{i,k}} \right\|} & \text{for } \left\| u_{eq_{i,k}} \right\| > u_{0i} \end{cases} \tag{3.31}$$

where the equivalent control $u_{eq_{i,k}}$ is calculated from $S_{D_{i,k+1}} = 0$ as

$$u_{eq_{i,k}} = \frac{1}{w_{i}^{'r}} \left(-w_{i,k}^{r} S(\chi_{i,k}^{1}, \dots, \chi_{i,k}^{r}) + x_{id,k+1}^{r} \right). \tag{3.32}$$

The whole proposed identification and control scheme for the system is displayed in Fig. 3.1.

Remark 3.4 ([13]) In fact, the system model must be expressed in the Nonlinear Block Controllable (NBC) form before to start the design. The robot manipulators considered in this book are already in this form [11, 12, 16].

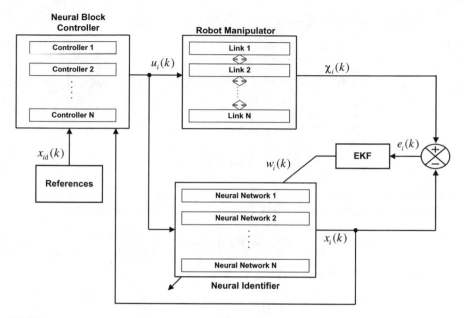

Fig. 3.1 Decentralized neural block control scheme

3.5 Stability Analysis

The stability analysis, to prove that the closed-loop system motion over the surface $S_{D_{i,k}}$ is stable, is stated as the following theorem.

Theorem 3.5 ([13]) *The control law (3.31) ensures the sliding manifold $S_{D_{i,k}} = \mathbf{z}_{i,k}^r = 0$ is stable for system (3.3).*

Proof Let write $S_{D_{i,k+1}}$ as

$$S_{D_{i,k+1}} = S_{D_{i,k}} - x_{i,k}^r + x_{id,k}^r + w_{i,k}^r S_i(\chi_{i,k}^1, \ldots, \chi_{i,k}^r)$$
$$+ w_i^{'r} u_{i,k} - x_{id,k+1}^r$$

Note that when $\|u_{eq_{i,k}}\| \leq u_{0_i}$, the equivalent control is applied, yielding motion on the sliding manifold $S_{D_{i,k}} = 0$. For $\|u_{eq_{i,k}}\| > u_{0_i}$, the proposed control strategy is $u_{0_i} \frac{u_{eq_{i,k}}}{\|u_{eq_{i,k}}\|}$, and the closed-loop system becomes

$$S_{D_{i,k+1}} = S_{D_{i,k}} - x_{i,k}^r + x_{id,k}^r + w_{i,k}^r S_i(\chi_{i,k}^1, \ldots, \chi_{i,k}^r)$$
$$+ w_i'^r u_{0_i} \frac{u_{eq_{i,k}}}{\|u_{eq_{i,k}}\|} - x_{id,k+1}$$
$$= \left(S_{D_{i,k}} - x_{i,k}^r + x_{id,k}^r + w_{i,k}^r S_i(\chi_{i,k}^1, \ldots, \chi_{i,k}^r) - x_{id,k+1}^r \right)$$
$$\left(1 - \frac{u_{0_i}}{\|u_{eq_{i,k}}\|} \right)$$
$$= (S_{D_{i,k}} + f_{s_{i,k}}) \left(1 - \frac{u_{0_i}}{\|u_{eq_{i,k}}\|} \right)$$

where $f_{s_{i,k}} = -x_{i,k}^r + x_{d,k}^r + w_{i,k}^r S_i(\chi_{i,k}^1, \ldots, \chi_{i,k}^r) - x_{d,k+1}^r$.

Along any solution of the system, the Lyapunov difference defined as

$$\Delta V_{i,k} = V_{i,k+1} - V_{i,k}$$

with $V_{i,k} = \|S_{D_{i,k}}\|$ becomes

$$\Delta V_{i,k} = \|S_{D_{i,k+1}}\| - \|S_{D_{i,k}}\|$$
$$= \|S_{D_{i,k}} + f_{s_{i,k}}\| \left(1 - \frac{u_{0_i}}{\|u_{eq_{i,k}}\|} \right) - \|S_{D_{i,k}}\|$$
$$= \|S_{D_{i,k}} + f_{s_{i,k}}\| \left(1 - \frac{u_{0_i}}{\|\frac{1}{w_i'^r}\| \|S_{D_{i,k}} + f_{s_{i,k}}\|} \right) - \|S_{D_{i,k}}\|$$
$$= \|S_{D_{i,k}} + f_{s_{i,k}}\| - \frac{u_{0_i}}{\|\frac{1}{w_i'^r}\|} - \|S_{D_{i,k}}\|$$
$$\leq \|S_{D_{i,k}}\| + \|f_{s_{i,k}}\| - \frac{u_{0_i}}{\|\frac{1}{w_i'^r}\|} - \|S_{D_{i,k}}\|$$
$$\leq \|f_{s_{i,k}}\| - \frac{u_{0_i}}{\|\frac{1}{w_i'^r}\|}$$

In [17] it is shown that under the condition $u_{0_i} > \|\frac{1}{w_i'^r}\| \|f_{s_{i,k}}\|$ the state vector of closed-loop system reaches the sliding manifold $S_{D_{i,k}}$ in finite time. Then the sliding motion $S_{D_{i,k}} = 0$ is governed by the following reduced order system (*sliding mode equation*, SME)

$$\mathbf{z}_{i,k+1}^1 = k_i^1 \mathbf{z}_{i,k}^1 + w_i'^1 \mathbf{z}_{i,k}^2$$
$$\mathbf{z}_{i,k+1}^2 = k_i^2 \mathbf{z}_{i,k}^2 + w_i'^2 \mathbf{z}_{i,k}^3$$
$$\vdots$$
$$\mathbf{z}_{i,k+1}^{r-1} = k_i^{r-1} \mathbf{z}_{i,k}^{r-1}$$

(3.33)

and if $|k_i^j| < 1$ the system (3.33) is asymptotically stable.

Lemma 3.6 *If Requirement 3.2 holds and $u_{0_i} > \|\frac{1}{w_i^{\prime r}}\| \| f_{s_{i,k}} \|$ is fulfilled, then the Lyapunov function candidate for the whole system defined as $V_k = \sum_{i=1}^{N} V_{i,k}$ whose difference $\Delta V_k < 0$ is negative definite, guarantees the asymptotic stability for the interconnected system.*

Proof Let $V_k = \sum_{i=1}^{N} V_{i,k}$, then

$$
\begin{aligned}
\Delta V_k &= \sum_{i=1}^{N} \left(\| S_{D_{i,k+1}} \| - \| S_{D_{i,k}} \| \right) \\
&= \sum_{i=1}^{N} \left(\| S_{D_{i,k}} + f_{s_{i,k}} \| \left(1 - \frac{u_{0_i}}{\| u_{eq_{i,k}} \|} \right) - \| S_{D_{i,k}} \| \right) \\
&= \sum_{i=1}^{N} \left(\| S_{D_{i,k}} + f_{s_{i,k}} \| - \frac{u_{0_i}}{\| \frac{1}{w_i^{\prime r}} \|} - \| S_{D_{i,k}} \| \right) \\
&\leq \sum_{i=1}^{N} \left(f_{s_{i,k}} \| - \frac{u_{0_i}}{\| \frac{1}{w_i^{\prime r}} \|} \right)
\end{aligned}
$$

and if $u_{0_i} > \| \frac{1}{w_i^{\prime r}} \| \| f_{s_{i,k}} \|$ holds, then $\Delta V_k < 0$.

References

1. Alanis, A.Y.: Discrete-Time Neural Control: Application to Induction Motors. Ph.D. thesis, Cinvestav, Unidad Guadalajara, Guadalajara, Jalisco, Mexico (2007)
2. Alanis, A.Y., Sanchez, E.N., Loukianov, A.G., Chen, G.: Discrete-time output trajectory tracking by recurrent high-order neural network control. In: Proceedings of the 45th IEEE Conference on Decision and Control, pp. 6367–6372, San Diego, CA, USA (2006)
3. Castañeda, C.E.: DC Motor Control based on Recurrent Neural Networks. Ph.D. thesis, Cinvestav, Unidad Guadalajara, Guadalajara, Jalisco, Mexico (2009)
4. Chui, C.K., Chen, G.: Kalman Filtering with Real-Time Applications. Springer, Berlin (2009)
5. Felix, R.A.: Variable Structure Neural Control. Ph.D. thesis, Cinvestav, Unidad Guadalajara, Guadalajara, Jalisco, Mexico (2003)
6. Felix, R.A., Sanchez, E.N., Loukianov, A.G.: Avoiding controller singularities in adaptive recurrent neural control. In: Proceedings of the 16th IFAC World Congress, pp. 1095–1100, Czech Republic, Prague (2005)
7. Grover, R., Hwang, P.Y.C.: Introduction to Random Signals and Applied Kalman Filtering. Wiley, New York (1992)
8. Haykin, S.: Kalman Filtering and Neural Networks. Wiley, New York (2001)
9. Ioannou, P., Sun, J.: Robust Adaptive Control. Prentice Hall Inc, Upper Saddle River (1996)
10. Lin, W., Byrnes, C.I.: Design of discrete-time nonlinear control systems via smooth feedback. IEEE Trans. Autom. Control **39**(11), 2340–2346 (1994)
11. Loukianov, A.G.: A block method of synthesis of nonlinear systems at sliding modes. Autom. Remote Control **59**(7), 916–933 (1998)
12. Loukianov, A.G.: Robust block decomposition sliding mode control design. Math. Probl. Eng. **8**(4–5), 349–365 (2003)

13. Sanchez, E.N., Alanis, A.Y., Loukianov, A.G.: Discrete-time high order neural control: trained with Kalman filtering. Springer, Berlin (2008)
14. Song, Y., Grizzle, J.W.: The extended kalman filter as a local asymptotic observer for discrete-time nonlinear systems. J. Math. Syst. Estim. Control 5(1), 59–78 (1995)
15. Utkin, V.: Sliding mode control design principles and applications to electric drives. IEEE Trans. Ind. Electron. 40(1), 23–36 (1993)
16. Utkin, V.: Block control principle for mechanical systems. J. Dyn. Syst. Meas. Contr. 122(1), 1–10 (2000)
17. Utkin, V., Guldner, J., Shi, J.: Sliding Mode Control in Electromechanical Systems. Taylor & Francis, Philadelphia (1999)

Chapter 4
Decentralized Neural Backstepping Control

4.1 Decentralized Systems in Block Strict Feedback Form

The model of many practical nonlinear systems can be expressed in (or transformed into) a special state form named the block strict feedback form (BSFF) [4] as follows

$$
\begin{aligned}
x_{i,k+1}^1 &= f_i^1\left(x_i^1\right) + g_i^1\left(x_i^1\right)x_{i,k}^2 + d_{i,k}^1 \\
x_{i,k+1}^2 &= f_i^2\left(x_i^1, x_i^2\right) + g_i^2\left(x_i^1, x_i^2\right)x_{i,k}^3 + d_{i,k}^2 \\
&\;\;\vdots \\
x_{i,k+1}^j &= f_i^j\left(x_i^1, x_i^2, \ldots, x_i^j\right) + g_i^j\left(x_i^1, x_i^2, \ldots, x_i^j\right)x_{i,k}^{j+1} + d_{i,k}^j \qquad (4.1)\\
&\;\;\vdots \\
x_{i,k+1}^r &= f_i^r\left(x_i\right) + g_i^r\left(x_i\right)u_{i,k} + d_{i,k}^r \\
y_{i,k} &= x_{i,k}^1
\end{aligned}
$$

where $x_i = \left[x_i^{1\top}\, x_i^{2\top}\, \ldots\, x_i^{r\top}\right]^\top$ are the state variables, $i = 1, \ldots, N$; $j = 1, \ldots, r-1$; N is the number of subsystems, r is the number of blocks, $r \geq 2$, $u_i \in \mathfrak{R}^{m_i}$ are the system inputs for each subsystem, $y_i \in \mathfrak{R}^{m_i}$ are the system outputs, $d_{i,k}^j \in \mathfrak{R}^{n_i}$ is the bounded disturbance vector which includes all the effects of the others connected systems, then there exists a constant $\overline{d}_i^j$ such that $\parallel d_{i,k}^j \parallel \leq \overline{d}_i^j$, $\forall i = 1, \ldots, N$; $j = 1, \ldots, r-1$ and for $0 < k < \infty$, $f_i^j(\bullet)$ and $g_i^j(\bullet)$ are unknown smooth nonlinear functions. If we consider the original system (4.1) as a one-step predictor, we can transform it into an equivalent maximum r-step ahead one, which can predict the future states $x_{i,k+r}^1, x_{i,k+r-1}^2, \ldots, x_{i,k+1}^r$; then, the causality contradiction is avoided when the controller is constructed based on the maximum r-step ahead prediction by backstepping [2, 3]; hence, system (4.1) can be rewritten as

© Springer International Publishing Switzerland 2017
R. Garcia-Hernandez et al., *Decentralized Neural Control: Application to Robotics*,
Studies in Systems, Decision and Control 96, DOI 10.1007/978-3-319-53312-4_4

$$x_{i,k+r}^1 = F_i^1(x_i^1) + G_i^1(x_i^1)x_{i,k+r-1}^2$$
$$+d_{i,k+r}^1$$

$$\vdots$$

$$x_{i,k+2}^{r-1} = F_i^{r-1}(x_i^1, x_i^2, \ldots, x_i^j) + G_i^{r-1}(x_i^1, x_i^2, \ldots, x_i^j)$$
$$x_{i,k+1}^r + d_{i,k+2}^{r-1} \qquad (4.2)$$
$$x_{i,k+1}^r = F_i^r(x_i) + G_i^r(x_i)u_{i,k} + d_{i,k}^r$$
$$y_{i,k} = x_{i,k}^1$$

where $F_i^j(\bullet)$ and $G_i^j(\bullet)$ are unknown smooth functions of $f_i^j(x_i^1, x_i^2, \ldots, x_i^j)$ and $g_i^j(x_i^1, x_i^2, \ldots, x_i^j)$ respectively. For analysis convenience, let us define $1 \le j \le r - 1$.

4.2 Approximation by High Order Neural Networks

Considering the high order neural network (HONN) described by

$$\phi(w, z) = w^\top S(z)$$
$$S(z) = [s_1^\top(z), s_2^\top(z), \ldots, s_m^\top(z)]$$
$$s_i(z) = \left[\prod_{j \in I_1}[s(z_j)]^{d_j(i_1)} \cdots \prod_{j \in I_m}[s(z_j)]^{d_j(i_m)} \right]^\top \qquad (4.3)$$
$$i = 1, 2, \ldots, L$$

where $z = [z_1, z_2, \ldots, z_p]^\top \in \Omega_z \subset \Re^p$, p is a positive integer which denotes the number of external inputs, L denotes the NN node number, $\phi \in \Re^m$, $\{I_1, I_2, \ldots, I_L\}$ is a collection of not ordered subsets of $\{1, 2, \ldots, p\}$, $S(z) \in \Re^{L \times m}$, $d_j(i_j)$ is a nonnegative integer, $w \in \Re^L$ is an adjustable synaptic weight vector, and $s(z_j)$ is selected as the hyperbolic tangent function:

$$s(z_j) = \frac{e^{z_j} - e^{-z_j}}{e^{z_j} + e^{-z_j}} \qquad (4.4)$$

For a desired function $u^* \in \Re^m$, let us assume there exists an ideal weight vector $w^* \in \Re^L$ such that the smooth function vector $u^*(z)$ can be approximated by an ideal NN on a compact subset $\Omega_z \subset \Re^q$

$$u^*(z) = w^{*\top} S(z) + \varepsilon_z \qquad (4.5)$$

where $\varepsilon_z \subset \Re^m$ is the bounded NN approximation error vector; note that $\|\varepsilon_z\|$ can be reduced by increasing the number of the adjustable weights. The ideal weight vector w^* is an artificial quantity required only for analytical purposes [3, 5]. In general, it is assumed that there exists an unknown but constant weight vector w^*, whose estimate is $w \in \Re^L$ Hence, it is possible to define:

$$\tilde{w}_k = w_k - w^* \tag{4.6}$$

as the weight estimation error.

4.3 Controller Design

Once each subsystem in the BSFF is defined, we apply the well-known backstepping technique [4]. We can define the desired virtual controls ($\alpha_{i,k}^{j*}$, $i = 1, \ldots, N$; $j = 1, \ldots, r-1$) and the ideal practical control ($u_{i,k}^*$) as follows:

$$
\begin{aligned}
\alpha_{i,k}^{1*} &\triangleq x_{i,k}^2 = \varphi_i^1(\bar{x}_{i,k}^1, x_{id,k+r}) \\
\alpha_{i,k}^{2*} &\triangleq x_{i,k}^3 = \varphi_i^2(\bar{x}_{i,k}^2, \alpha_{i,k}^{1*}) \\
&\vdots \\
\alpha_{i,k}^{r-1*} &\triangleq x_{i,k}^r = \varphi_i^{r-1}(\bar{x}_{i,k}^{r-1}, \alpha_{i,k}^{r-2*}) \\
u_{i,k}^* &= \varphi_i^r(x_{i,k}, \alpha_{i,k}^{r-1*}) \\
\chi_{i,k} &= x_{i,k}^1
\end{aligned}
\tag{4.7}
$$

where $\varphi_i^j(\bullet)$ with $1 \le j \le r$ are nonlinear smooth functions. It is obvious that the desired virtual controls $\alpha_{i,k}^{j*}$ and the ideal control $u_{i,k}^*$ will drive the output $\chi_{i,k}$ to track the desired signal $x_{id,k}$ only if the exact system model is known and there are no unknown disturbances; however in practical applications these two conditions cannot be satisfied. In the following, neural networks will be used to approximate the desired virtual controls, as well as the desired practical controls, when the conditions established above are not satisfied. As in [3], we construct the virtual and practical controls via embedded backstepping without the causality contradiction [1, 2]. Let us approximate the virtual controls and the practical control by the following HONN:

$$
\begin{aligned}
\alpha_{i,k}^j &= w_i^{jT} S_i^j(z_{i,k}^j), \quad i = 1, \ldots, N \\
u_{i,k} &= w_i^{rT} S_i^r(z_{i,k}^r), \quad j = 1, \ldots, r-1
\end{aligned}
\tag{4.8}
$$

with

$$z_{i,k}^1 = [x_{i,k}^1, x_{id,k+r}^1]^\top, \quad i = 1, \dots, N$$
$$z_{i,k}^j = [\overline{x}_{i,k}^j, \alpha_{i,k}^{j-1}]^\top, \quad j = 1, \dots, r-1$$
$$z_{i,k}^r = [x_{i,k}, \alpha_{i,k}^{r-1}]^\top$$

where $w_i^j \in \Re^{L_{ij}}$ are the estimates of ideal constant weights w_i^{j*} and $S_i^j \in \Re^{L_{ij} \times n_{ij}}$ with $j = 1, \dots, r$. Define the weight estimation error as

$$\tilde{w}_{i,k}^j = w_{i,k}^j - w_i^{j*}. \tag{4.9}$$

Using the ideal constant weights and from (4.5), it follows that exist a HONN, which approximate the virtual controls and practical control with a minimum error, defined as:

$$\alpha_{i,k}^j = w_i^{j*^\top} S_i^j(z_{i,k}^j), \quad i = 1, \dots, N$$
$$u_{i,k} = w_i^{r*^\top} S_i^r(z_{i,k}^r) + \varepsilon_{z_i^j}, \quad j = 1, \dots, r-1 \tag{4.10}$$

Then, the corresponding weights updating laws are defined as

$$w_{i,k+1}^j = w_{i,k}^j + \eta_i^j K_{i,k}^j e_{i,k}^j \tag{4.11}$$

with

$$K_{i,k}^j = P_{i,k}^j H_{i,k}^j M_{i,k}^j$$
$$M_{i,k}^j = [R_{i,k}^j + H_{i,k}^{j\top} P_{i,k}^j H_{i,k}^j]^{-1} \tag{4.12}$$
$$P_{i,k+1}^j = P_{i,k}^j - K_{i,k}^j H_{i,k}^{j\top} P_{i,k}^j + Q_{i,k}^j$$

$$H_{i,k}^j = \left[\frac{\partial \hat{v}_{i,k}^j}{\partial w_{i,k}^j} \right] \tag{4.13}$$

and

$$e_{i,k}^j = v_{i,k}^j - \hat{v}_{i,k}^j \tag{4.14}$$

where $v_{i,k}^j \in \Re^{n_{ij}}$ is the desired reference to be tracked and $\hat{v}_{i,k}^j \in \Re^{n_{ij}}$ is the HONN function approximation defined, respectively as follows

$$v_{i,k}^1 = x_{id,k}^1$$
$$v_{i,k}^2 = x_{i,k}^2$$
$$\vdots$$
$$v_{i,k}^r = x_{i,k}^r \tag{4.15}$$

and

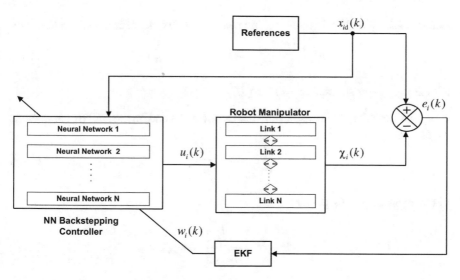

Fig. 4.1 Decentralized neural backstepping control scheme

$$\hat{\upsilon}_{i,k}^1 = \chi_{i,k}^1$$
$$\hat{\upsilon}_{i,k}^2 = \alpha_{i,k}^1$$
$$\vdots \tag{4.16}$$
$$\hat{\upsilon}_{i,k}^r = \alpha_{i,k}^{r-1}$$

$e_{i,k}^j$ denotes the error at each step as

$$e_{i,k}^1 = x_{\mathrm{id},k}^1 - \chi_{i,k}^1$$
$$e_{i,k}^2 = x_{i,k}^2 - \alpha_{i,k}^1$$
$$\vdots \tag{4.17}$$
$$e_{i,k}^r = x_{i,k}^r - \alpha_{i,k}^{r-1}.$$

The whole proposed decentralized neural backstepping control scheme is shown in Fig. 4.1.

4.4 Stability Analysis

Before proceeding to demonstrate the stability analysis to prove that the tracking error (4.17) is SGUUB, we need to establish the following lemmas.

Lemma 4.1 ([6]) *The dynamics of the tracking errors (4.14) can be formulated as*

$$e_{i,k+1}^j = e_{i,k}^j + \Delta e_{i,k}^j, \quad (1 \le j \le r), \tag{4.18}$$

with $\Delta e_{i,k}^j \le -\gamma_i^j e_{i,k}^j$ and $\gamma_i^j = \max \| H_{i,k}^{j\top} \eta_i^j K_{i,k}^j \|$.

Proof Using (4.14) and considering that υ_k do not depend on the HONN parameters, we obtain

$$\frac{\partial e_{i,k}^j}{\partial w_{i,k}^j} = -\frac{\partial \hat{\upsilon}_{i,k}^j}{\partial w_{i,k}^j} \tag{4.19}$$

Let us approximate (4.19) by

$$\Delta e_{i,k}^j = \left[\frac{\partial e_{i,k}^j}{\partial w_{i,k}^j} \right]^\top \Delta w_{i,k}^j. \tag{4.20}$$

Substituting (4.13) and (4.19) in (4.20) yields

$$\Delta e_{i,k}^j = -H_{i,k}^{j\top} \eta_i^j K_{i,k}^j e_i^j. \tag{4.21}$$

Let define

$$\gamma_i^j = \max \| H_{i,k}^{j\top} \eta_i^j K_{i,k}^j \| \tag{4.22}$$

then we have

$$\Delta e_{i,k}^j = -\gamma_i^j e_{i,k}^j. \tag{4.23}$$

Lemma 4.2 ([6]) *The HONN weights updated with (4.11), based on the EKF algorithm (4.12), are bounded.*

Proof From (4.9) and (4.11) it is possible to write the dynamics of the weight estimation error as

$$\tilde{w}_{i,k+1}^j = \tilde{w}_{i,k}^j + \eta_i^j K_{i,k}^j e_{i,k}^j \tag{4.24}$$

Using (4.8), (4.10), and (4.13) the system (4.24), can be written as

$$\begin{aligned}
\tilde{w}_{i,k+1}^j &= \tilde{w}_{i,k}^j - \eta_i^j K_i^j S^{j\top}(z_{i,k}^j) \tilde{w}_{i,k}^j + \eta_i^j K_{i,k}^j \varepsilon_{z_i^j} \\
&= A_{i,k}^j \tilde{w}_{i,k}^1 + B_i^j \upsilon_{z_{i,k}^j}, \quad j = 1, \dots, r
\end{aligned}$$

with

$$\begin{aligned}
A_{i,k}^j &= [I - \eta_i^j K_{i,k}^j S^{j\top}(z_{i,k}^j)] \\
B_{i,k}^j &= \eta_i^j \\
\upsilon_{z_{i,k}^j} &= K_{i,k}^j \varepsilon_{z_i^j}
\end{aligned} \tag{4.25}$$

Considering Remarks 2.9 and 2.10, and the boundedness of $\varepsilon_{z_i^j}$ and $S(z_{i,k}^j)$, then, by selecting η_i^j appropriately, $A_{i,k}^j$ satisfies $\|\Phi(k(1), k(0))\| < 1$. By applying Lemma 2.6, $\tilde{w}_{i,k}^j$ is bounded.

Theorem 4.3 *For the i-th subsystem of (4.1) in absence of interconnections, the i-th subsystem of HONN (4.8) trained with the EKF-based algorithm (4.12) to approximate the i-th control law (4.7), ensures that the tracking error (4.17) is semiglobally uniformly ultimately bounded (SGUUB); moreover, the HONN weights remain bounded.*

Proof For the first block of i-th subsystem in (4.1), with the virtual control $\alpha_{i,k}^{1*}$ approximated by the i-th subsystem of HONN ($\alpha_{i,k}^1 = w_i^{1\top} S_i^1(z_{i,k}^1)$) and $e_{i,k}^1$ defined as in (4.17), consider the Lyapunov function candidate

$$V_{i,k}^1 = e_{i,k}^{1\top} e_{i,k}^1 + \tilde{w}_{i,k}^{1\top} \tilde{w}_{i,k}^1 \tag{4.26}$$

whose first difference is

$$\begin{aligned} \Delta V_{i,k}^1 &= V_{i,k+1}^1 - V_{i,k}^1 \\ &= e_{i,k+1}^{1\top} e_{i,k+1}^1 + \tilde{w}_{i,k+1}^{1\top} \tilde{w}_{i,k+1}^1 - e_{i,k}^{1\top} e_{i,k}^1 \\ &\quad + \tilde{w}_{i,k}^{1\top} \tilde{w}_{i,k}^1 \end{aligned} \tag{4.27}$$

From (4.9) and (4.11)

$$\tilde{w}_{i,k+1}^1 = \tilde{w}_{i,k}^1 + \eta_i^1 K_{i,k}^1 e_{i,k}^1 \tag{4.28}$$

Let us define

$$\begin{aligned} [\tilde{w}_{i,k}^1 + \eta_i^1 K_{i,k}^1 e_{i,k}^1]^\top [\tilde{w}_{i,k}^1 &+ \eta_i^1 K_{i,k}^1 e_{i,k}^1] \\ &= \tilde{w}_{i,k}^{1\top} \tilde{w}_{i,k}^1 + 2\tilde{w}_{i,k}^{1\top} \eta_i^1 K_{i,k}^1 e_{i,k}^1 + (\eta_i^1 K_{i,k}^1 e_{i,k}^1)^\top \eta_i^1 K_{i,k}^1 e_{i,k}^1 \end{aligned} \tag{4.29}$$

From (4.17), then

$$\begin{aligned} e_{i,k+1}^1 &= e_{i,k+1}^1 + \Delta e_{i,k}^1 \\ e_{i,k+1}^{1\top} e_{i,k+1}^1 &= e_{i,k}^{1\top} e_{i,k}^1 + e_{i,k}^{1\top} \Delta e_{i,k}^1 + \Delta e_{i,k}^{1\top} e_{i,k}^1 \\ &\quad + \Delta e_{i,k}^{1\top} \Delta e_{i,k}^1 \\ e_{i,k+1}^{1\top} e_{i,k+1}^1 - e_{i,k}^{1\top} e_{i,k}^1 &= e_{i,k}^{1\top} \Delta e_{i,k}^1 + \Delta e_{i,k}^{1\top} e_{i,k}^1 \\ &\quad + \Delta e_{i,k}^{1\top} \Delta e_{i,k}^1 \end{aligned}$$

where $\Delta e_{i,k}^1$ is the error difference. Substituting (4.28) and (4.29) in (4.27) results in

$$\begin{aligned} \Delta V_{i,k}^1 &= e_{i,k}^{1\top} \Delta e_{i,k}^1 + \Delta e_{i,k}^{1\top} e_{i,k}^1 + \Delta e_{i,k}^{1\top} \Delta e_{i,k}^1 \\ &\quad + 2\tilde{w}_{i,k}^{1\top} \eta_i^1 K_{i,k}^1 e_{i,k}^1 + (\eta_i^1 K_{i,k}^1 e_{i,k}^1)^{1\top} \eta_i^1 K_{i,k}^1 e_{i,k}^1 \end{aligned} \tag{4.30}$$

From Lemma 4.1, substituting (4.23), we obtain

$$
\begin{aligned}
\Delta V_{i,k}^1 &\leq -2\gamma_i^1 e_{i,k}^{1\mathrm{T}} e_{i,k}^1 + \gamma_1^{1^2} e_{i,k}^{1\mathrm{T}} e_{i,k}^1 + 2\tilde{w}_{i,k}^{1\mathrm{T}} \eta_i^1 K_{i,k}^1 e_{i,k}^1 \\
&\quad + (\eta_i^1 K_{i,k}^1 e_{i,k})^{1\mathrm{T}} \eta_i^1 K_{i,k}^1 e_{i,k}^1 \\
&\leq -2\gamma_i^1 \|e_{1,k}\|^2 + \gamma_i^{1^2} \|e_{1,k}\|^2 + 2\|\tilde{w}_{i,k}^{1\mathrm{T}} \eta_i^1 K_{i,k}^1\| \|e_{i,k}^1\| \\
&\quad + \|(\eta_i^1 K_{i,k}^1\|^2 \|e_{i,k}^1\|^2
\end{aligned}
\tag{4.31}
$$

where $\gamma_i^1 = \max \|H_{i,k}^{1\mathrm{T}} \eta_i^1 K_{i,k}^1\|$. From Lemma 4.2, it follows that $\tilde{w}_{i,k}^1$ is bounded; then, there is $\eta_i^1 > 0$ such that

$$
\Delta V_{i,k}^1 \leq 0, \quad \text{once } \|e_{i,k}^1\| > \kappa_i^1
\tag{4.32}
$$

with κ_i^1 defined as

$$
\kappa_i^1 = \frac{2\eta_i^1 \bar{w}_i^1 \bar{K}_i^1}{2\gamma_i^1 - \gamma_i^{1^2} - \eta_i^{1^2} \bar{K}_i^{1^2}}
$$

where $\bar{w}_i^1$ and $\bar{K}_i^1$ are the upper bound of $\tilde{w}_{i,k}^1$ and $K_{i,k}^1$, respectively [7]. From (4.32), it follows the boundedness of $V_{i,k}^1$ for $k \geq k_T$, that leads to the SGUUB of $e_{i,k}^1$.

For the following j-th ($j = 2, \ldots, r - 1$) equations of i-th subsystem in (4.1), with the virtual control $\alpha_{i,k}^{j*}$ approximated by the i-th subsystem of HONN ($\alpha_{i,k}^j = w_i^{j\mathrm{T}} S_i^j (z_{i,k}^j)$) and $e_{i,k}^j$ defined as in (4.16), consider the Lyapunov function candidate

$$
V_{i,k}^j = e_{i,k}^{j\mathrm{T}} e_{i,k}^j + \tilde{w}_{i,k}^{j\mathrm{T}} \tilde{w}_{i,k}^j
\tag{4.33}
$$

whose first difference is

$$
\begin{aligned}
\Delta V_{i,k}^j &= V_{i,k+1}^j - V_{i,k}^j \\
&= e_{i,k+1}^{j\mathrm{T}} e_{i,k+1}^j + \tilde{w}_{i,k+1}^{j\mathrm{T}} \tilde{w}_{i,k+1}^j - e_{i,k}^{j\mathrm{T}} e_{i,k}^j \\
&\quad + \tilde{w}_{i,k}^{j\mathrm{T}} \tilde{w}_{i,k}^j
\end{aligned}
\tag{4.34}
$$

From (4.9) and (4.11)

$$
\tilde{w}_{i,k+1}^j = \tilde{w}_{i,k}^j + \eta_i^j K_{i,k}^j e_{i,k}^j
\tag{4.35}
$$

Let us define

$$
\begin{aligned}
[\tilde{w}_{i,k}^j + \eta_i^j K_{i,k}^j e_{i,k}^j]^{\mathrm{T}} [\tilde{w}_{i,k}^j + \eta_i^j K_{i,k}^j e_{i,k}^j] \\
= \tilde{w}_{i,k}^{j\mathrm{T}} \tilde{w}_{i,k}^j + 2\tilde{w}_{i,k}^{j\mathrm{T}} \eta_i^j K_{i,k}^j e_{i,k}^j + (\eta_i^j K_{i,k}^j e_{i,k}^j)^{\mathrm{T}} \eta_i^j K_{i,k}^j e_{i,k}^j
\end{aligned}
\tag{4.36}
$$

From (4.17), then

$$e_{i,k+1}^j = e_{i,k+1}^j + \Delta e_{i,k}^j$$

$$e_{i,k+1}^{j\top} e_{i,k+1}^j = e_{i,k}^{j\top} e_{i,k}^j + e_{i,k}^{j\top} \Delta e_{i,k}^j + \Delta e_{i,k}^{j\top} e_{i,k}^j$$
$$+ \Delta e_{i,k}^{j\top} \Delta e_{i,k}^j$$

$$e_{i,k+1}^{j\top} e_{i,k+1}^j - e_{i,k}^{j\top} e_{i,k}^j = e_{i,k}^{j\top} \Delta e_{i,k}^j + \Delta e_{i,k}^{j\top} e_{i,k}^j$$
$$+ \Delta e_{i,k}^{j\top} \Delta e_{i,k}^j$$

where $\Delta e_{i,k}^j$ is the error difference. Substituting (4.35) and (4.36) in (4.34) results in

$$\Delta V_{i,k}^j = e_{i,k}^{j\top} \Delta e_{i,k}^j + \Delta e_{i,k}^{j\top} e_{i,k}^j + \Delta e_{i,k}^{j\top} \Delta e_{i,k}^j$$
$$+ 2\tilde{w}_{i,k}^{j\top} \eta_i^j K_{i,k}^j e_{i,k}^j + (\eta_i^j K_{i,k}^j e_{i,k}^j)^\top \eta_i^j K_{i,k}^j e_{i,k}^j \qquad (4.37)$$

From Lemma 4.1, substituting (4.23), we obtain

$$\Delta V_{i,k}^j \leq -2\gamma_i^j e_{i,k}^{j\top} e_{i,k}^j + \gamma_i^{j^2} e_{i,k}^{1\top} e_{i,k}^j + 2\tilde{w}_{i,k}^{j\top} \eta_i^j K_{i,k}^j e_{i,k}^j$$
$$+ (\eta_i^j K_{i,k}^j e_{i,k}^j)^\top \eta_i^j K_{i,k}^j e_{i,k}^j$$
$$\leq -2\gamma_i^j \|e_{i,k}^j\|^2 + \gamma_i^{j^2} \|e_{i,k}^j\|^2 + 2\|\tilde{w}_{i,k}^{j\top} \eta_i^j K_{i,k}^j\| \|e_{i,k}^j\| \qquad (4.38)$$
$$+ \|(\eta_i^j K_{i,k}^j\|^2 \|e_{i,k}^j\|^2$$

where $\gamma_i^j = \max \|H_{i,k}^{j\top} \eta_i^j K_{i,k}^j\|$. From Lemma 4.2, it follows that $\tilde{w}_{i,k}^j$ is bounded; then, there is $\eta_i^j > 0$ such that

$$\Delta V_{i,k}^j \leq 0, \quad \text{once } \|e_{i,k}^j\| > \kappa_i^j \qquad (4.39)$$

with κ_i^j defined as

$$\kappa_i^j = \frac{2\eta_i^j \bar{w}_i^j \bar{K}_i^j}{2\gamma_i^j - \gamma_i^{j^2} - \eta_i^{j^2} \bar{K}_i^{j^2}}$$

where $\bar{w}_i^j$ and $\bar{K}_i^j$ are the upper bound of $\tilde{w}_{i,k}^j$ and $K_{i,k}^j$, respectively [7]. From (4.39), it follows the boundedness of $V_{i,k}^j$ for $k \geq k_T$, that leads to the SGUUB of $e_{i,k}^j$.

Theorem 4.4 *For the i-th subsystem of (4.1) in presence of interconnections, the i-th subsystem of HONN (4.8) with $i = 1, \ldots, N$; $j = 1, \ldots, r_i$ trained with the EKF-based algorithm (4.12) to approximate the i-th control law (4.7), ensures that the tracking error (4.17) is semiglobally uniformly ultimately bounded (SGUUB); moreover, the HONN weights remain bounded.*

Proof Let $V_k = \sum_{i=1}^{N} \sum_{j=1}^{r_i} V_{i,k}^j$, then

$$\Delta V_k = \sum_{i=1}^{N} \sum_{j=1}^{r_i} \left(e_{i,k}^{j\top} \Delta e_{i,k}^j + \Delta e_{i,k}^{j\top} e_{i,k}^j + \Delta e_{i,k}^{j\top} \Delta e_{i,k}^j \right.$$
$$\left. + 2\tilde{w}_{i,k}^{j\top} \eta_i^j K_{i,k}^j e_{i,k}^j + (\eta_i^j K_{i,k}^j e_{i,k}^j)^\top \eta_i^j K_{i,k}^j e_{i,k}^j \right)$$

substituting $\Delta e_{i,k}^j \leq -\gamma_i^j e_{i,k}^j$, we obtain

$$\Delta V_k \leq \sum_{i=1}^{N} \sum_{j=1}^{r_i} \left(-2\gamma_i^j e_{i,k}^{j\top} e_{i,k}^j + \gamma_1^{j^2} e_{i,k}^{j\top} e_{i,k}^j \right.$$
$$\left. + 2\tilde{w}_{i,k}^{j\top} \eta_i^j K_{i,k}^j e_{i,k}^j + (\eta_i^j K_{i,k}^j e_{i,k}^j)^\top \eta_i^j K_{i,k}^j e_{i,k}^j \right)$$
$$\leq \sum_{i=1}^{N} \sum_{j=1}^{r_i} \left(-2\gamma_1^j \|e_{i,k}^j\|^2 + \gamma_1^{j^2} \|e_{i,k}^j\|^2 \right.$$
$$\left. + 2\|\tilde{w}_{i,k}^{j\top} \eta_i^j K_{i,k}^j\|\|e_{i,k}^j\| + \|(\eta_i^j K_{i,k}^j\|^2 \|e_{i,k}^j\|^2) \right)$$

where $\gamma_i^j = \max \|H_{i,k}^{j\top} \eta_i^j K_{i,k}^j\|$. From Lemma 4.2, it follows that $\tilde{w}_{i,k}^j$ is bounded; then, there is $\eta_i^j > 0$ such that

$$\Delta V_k \leq 0, \quad \text{once } \|e_{i,k}^j\| > \kappa_i^j \tag{4.40}$$

with κ_i^j defined as

$$\kappa_i^j = \frac{2\eta_i^j \bar{w}_i^j \bar{K}_i^j}{2\gamma_i^j - \gamma_i^{j^2} - \eta_i^{j^2} \bar{K}_i^{j^2}}$$

where $\bar{w}_i^j$ and $\bar{K}_i^j$ are the upper bound of $\tilde{w}_{i,k}^j$ and $K_{i,k}^j$, respectively [7]. From (4.40), it follows the boundedness of V_k for $k \geq k_T$, that leads to the SGUUB of $e_{i,k}^j$ $\forall i = 1, \ldots, N; j = 1, \ldots, r_i$.

References

1. Alanis, A.Y.: Discrete-Time Neural Control: Application to Induction Motors. Ph.D. thesis, Cinvestav, Unidad Guadalajara, Guadalajara, Jalisco, Mexico (2007)
2. Chen, F., Khalil, H.K.: Adaptive control of a class of nonlinear discrete-time systems using neural networks. IEEE Trans. Autom. Control **40**(5), 791–801 (1995)

3. Ge, S.S., Zhang, J., Lee, T.H.: Adaptive neural network control for a class of MIMO nonlinear systems with disturbances in discrete-time. IEEE Trans. Syst. Man Cybern. Part B Cybern. **34**(4), 1630–1645 (2004)
4. Krstic, M., Kanellakopoulos, I., Kokotovic, P.V.: Nonlinear and Adaptive Control Design. Wiley, New York (1995)
5. Rovithakis, G.A., Christodoulou, M.A.: Adaptive Control with Recurrent High-Order Neural Networks. Springer, Berlin (2000)
6. Sanchez, E.N., Alanis, A.Y., Loukianov, A.G.: Discrete-Time High Order Neural Control: Trained with Kalman Filtering. Springer, Berlin (2008)
7. Song, Y., Grizzle, J.W.: The extended kalman filter as a local asymptotic observer for discrete-time nonlinear systems. J. Math. Syst. Estim. Control **5**(1), 59–78 (1995)

Chapter 5
Decentralized Inverse Optimal Control for Stabilization: A CLF Approach

5.1 Decentralized Inverse Optimal Control via CLF

Let consider a class of disturbed discrete-time nonlinear and interconnected system

$$\chi_{i,k+1}^{j} = f_i^{j}(\chi_{i,k}^{j}) + g_i^{j}(\chi_{i,k}^{j})u_{i,k} + \Gamma_{i\ell,k}^{j}(\chi_\ell) \tag{5.1}$$

where $i = 1, \ldots, \gamma$, $j = 1, \ldots, n_i$, $\chi_i \in \Re^{n_i}, \chi_i = [\chi_i^{1^\top} \chi_i^{2^\top} \ldots \chi_i^{r^\top}]^\top$, $\chi_i^{j} \in \Re^{n_{ij} \times 1}$, $u_i \in \Re^{m_i}$, γ is the number of subsystems, $\Gamma_{i\ell}(\chi_\ell)$ reflect the interaction between the i-th and the ℓ-th subsystem with $1 \le \ell \le \gamma$. We assume that f_i, B_i and Γ_i are smooth and bounded functions, $f_i^{j}(0) = 0$ and $B_i^{j}(0) = 0$. Without loss of generality, $\chi = 0$ is an equilibrium point of (5.1), which is to be used later.

For the inverse optimal control approach, let consider the discrete-time affine in the input nonlinear system:

$$\chi_{i,k+1} = f_i(\chi_{i,k}) + g_i(\chi_{i,k})u_{i,k}, \quad \chi_{i,0} = \chi_i(0)$$
$$i = 1, \ldots, \gamma \tag{5.2}$$

where $\chi_{i,k} \in \Re^{n_i}$ are the states of the systems, $u_{i,k} \in \Re^{m_i}$ are the control inputs, $f_i(\chi_{i,k}) : \Re^{n_i} \rightarrow \Re^{n_i}$ and $g(\chi_k) : \Re^{n_i} \rightarrow \Re^{n_i \times m_i}$ are smooth maps, the subscript $k \in \mathbb{Z}^+ \cup 0 = \{0, 1, 2, \ldots\}$ stands for the value of the functions and/or variables at the time k. We establish the following assumptions and definitions which allow the inverse optimal control solution via the CLF approach.

Assumption 5.1 The full state of system (5.2) is measurable.

Definition 5.1 Let define the control law

$$u_{i,k}^{*} = -\frac{1}{2}R_i^{-1}(\chi_{i,k})g_i^{\top}(\chi_{i,k})\frac{\partial V_i(\chi_{i,k+1})}{\partial \chi_{i,k+1}} \tag{5.3}$$

to be inverse optimal (globally) stabilizing if:

© Springer International Publishing Switzerland 2017
R. Garcia-Hernandez et al., *Decentralized Neural Control: Application to Robotics*, Studies in Systems, Decision and Control 96, DOI 10.1007/978-3-319-53312-4_5

- (i) It achieves (global) asymptotic stability of $\chi_i = 0$ for system (5.2);
- (ii) $V_i(\chi_{i,k})$ is (radially unbounded) positive definite function such that inequality

$$\overline{V}_i := V_i(\chi_{i,k+1}) - V_i(\chi_{i,k}) + u_{i,k}^{*\top} R_i(\chi_{i,k}) u_{i,k}^* \leq 0 \tag{5.4}$$

is satisfied.

When $l_i(\chi_{i,k}) := -\overline{V}_i \leq 0$ is selected, then $V_i(\chi_{i,k})$ is a solution for the HJB equation

$$0 = l_i(\chi_{i,k}) + V_i(\chi_{i,k+1}) - V_i(\chi_{i,k}) \tag{5.5}$$
$$+ \frac{1}{4} \frac{\partial V_i^\top(\chi_{i,k+1})}{\partial \chi_{i,k+1}} g_i(\chi_{i,k}) R^{-1}(\chi_{i,k}) g_i^\top(\chi_{i,k}) \frac{\partial V_i(\chi_{i,k+1})}{\partial \chi_{i,k+1}}$$

It is possible to establish the main conceptual differences between optimal control and inverse optimal control as follows:

- For optimal control, the meaningful cost indexes $l_i(\chi_{i,k}) \leq 0$ and $R_i(\chi_{i,k}) > 0$ are given a priori; then, they are used to calculate $u_i(\chi_{i,k})$ and $V_i(\chi_{i,k})$ by means of the HJB equation solution.
- For inverse optimal control, a candidate CLF ($V_i(\chi_{i,k})$) and the meaningful cost index $R_i(\chi_{i,k})$ are given a priori, and then these functions are used to calculate the inverse control law $u_i^*(\chi_{i,k})$ and the meaningful cost index $l_i(\chi_{i,k})$, defined as $l_i(\chi_i) := -\overline{V}_i$.

As established in Definition 5.1, the inverse optimal control problem is based on the knowledge of $V_i(\chi_{i,k})$. Thus, we propose a CLF $V_i(\chi_{i,k})$, such that (i) and (ii) are guaranteed. That is, instead of solving (2.31) for $V_i(\chi_{i,k})$, we propose a control Lyapunov function $V_i(\chi_{i,k})$ with the form:

$$V_i(\chi_{i,k}) = \frac{1}{2} \chi_{i,k}^\top P_i \chi_{i,k}, \quad P_i = P_i^\top > 0 \tag{5.6}$$

for control law (5.3) in order to ensure stability of the equilibrium point $\chi_{i,k} = 0$ of system (5.2), which will be achieved by defining an appropriate matrix P_i.

Moreover, it will be established that control law (5.3) with (5.6), which is referred to as the inverse optimal control law, optimizes a meaningful cost function of the form (2.24).

Consequently, by considering $V_i(\chi_{i,k})$ as in (5.6), control law (5.3) takes the following form:

$$\alpha(\chi_{i,k}) := u_{i,k}^* = -\frac{1}{2} (R_i(\chi_{i,k}) + P_{i,2}(\chi_{i,k}))^{-1} P_{i,1}(\chi_{i,k}) \tag{5.7}$$

where $P_{i,1}(\chi_{i,k}) = g_i^\top(\chi_{i,k}) P_i f_i(\chi_{i,k})$ and $P_{i,2}(\chi_{i,k}) = \frac{1}{2} g_i^\top(\chi_{i,k}) P_i g_i(\chi_{i,k})$. It is worth to point out that P_i and $R_i(\chi_{i,k})$ are positive definite and symmetric matrices; thus, the existence of the inverse in (5.7) is ensured.

Once we have proposed a CLF for solving the inverse optimal control in accordance with Definition 5.1, the respective solution is presented, for which P_i is considered a fixed matrix.

Lemma 5.2 *Consider the affine discrete-time nonlinear system (5.2) with $i = 1$, If there exists a matrix $P_i = P_i^\top > 0$ such that the following inequality holds:*

$$V_{i,f}(\chi_{i,k}) - \frac{1}{4}P_{i,1}^\top(\chi_{i,k})(R_i(\chi_{i,k}) + P_{i,2}(\chi_{i,k}))^{-1}P_{i,1}(\chi_{i,k})$$

$$\leq -\zeta_{i,Q}\|\chi_{i,k}\|^2 \qquad (5.8)$$

where $V_{i,f}(\chi_{i,k}) = V_i(f_i(\chi_{i,k})) - V_i(\chi_{i,k})$, with $V_i(f_i(\chi_{i,k})) = \frac{1}{2}f_i^\top(\chi_{i,k})P_i f_i(\chi_{i,k})$ and $\zeta_{i,Q} > 0$; $P_{i,1}(\chi_{i,k})$ and $P_{i,2}(\chi_{i,k})$ as defined in (5.7); then, the equilibrium point $\chi_{i,k} = 0$ of system (5.2) is globally exponentially stabilized by the control law (5.7), with CLF (5.6).

Moreover, with (5.6) as a CLF, this control law is inverse optimal in the sense that it minimizes the meaningful functional given by

$$\mathscr{J}_i(\chi_{i,k}) = \sum_{k=0}^{\infty}(l_i(\chi_{i,k}) + u_{i,k}^\top R_i(\chi_{i,k})u_{i,k}) \qquad (5.9)$$

with

$$l_i(\chi_{i,k}) = -\overline{V}_i|_{u_{i,k}^* = \alpha_i(\chi_{i,k})} \qquad (5.10)$$

and optimal value function $\mathscr{J}_i(\chi_{i,k}) = V_i(\chi_0)$.

This lemma is adapted from [10] for each isolated subsystem, which allows to establish the following theorem.

Theorem 5.3 *Consider the affine discrete-time nonlinear system (5.2) with $i = 1, \ldots, \gamma$; γ the number of subsystems. If there exists matrices $P_i = P_i^\top > 0$ such that the following inequality holds:*

$$\sum_{i=0}^{\gamma}\left[V_{i,f}(\chi_{i,k}) - \frac{1}{4}P_{i,1}^\top(\chi_{i,k})(R_i(\chi_{i,k}) + P_{i,2}(\chi_{i,k}))^{-1}P_{i,1}(\chi_{i,k})\right] \qquad (5.11)$$

$$\leq -\sum_{i=0}^{\gamma}\zeta_{i,Q}\|\chi_{i,k}\|^2$$

where $\sum_{i=0}^{\gamma}[V_{i,f}(\chi_{i,k})] = \sum_{i=0}^{\gamma}[V_i(f_i(\chi_{i,k})) - V_i(\chi_{i,k})]$, with $\sum_{i=0}^{\gamma}\frac{1}{2}[f_i^\top(\chi_{i,k}) P_i f_i(\chi_{i,k})] = \sum_{i=0}^{\gamma}[V_i(f_i(\chi_{i,k}))]$ and $\sum_{i=0}^{\gamma}[\zeta_{i,Q}] > 0$; $P_{i,1}(\chi_{i,k})$ and $P_{i,2}(\chi_{i,k})$ as defined in (5.7); then, the equilibrium point $\chi_{i,k} = 0$ of system (5.2) is globally exponentially stabilized by the control law (5.7), with CLF (5.6).

Moreover, with (5.6) as a CLF, this control law is inverse optimal in the sense that it minimizes the meaningful functional given by

$$\sum_{i=0}^{\gamma} \mathscr{J}_i(\chi_{i,k}) = \sum_{i=0}^{\gamma} \sum_{k=0}^{\infty} (l_i(\chi_{i,k}) + u_{i,k}^{\top} R_i(\chi_{i,k}) u_{i,k}) \tag{5.12}$$

with

$$\sum_{i=0}^{\gamma} l_i(\chi_{i,k}) = -\sum_{i=0}^{\gamma} \overline{V}_i |_{u_{i,k}^* = \alpha_i(\chi_{i,k})} \tag{5.13}$$

and optimal value function $\sum_{i=0}^{\gamma}[\mathscr{J}_i(\chi_{i,k})] = \sum_{i=0}^{\gamma}[V_i(\chi_0)].$

Proof First, we analyze stability. Global stability for the equilibrium point $\sum_{i=0}^{\gamma}[\chi_{i,k}]$ = 0 of system (5.2), with (5.7) as input, is achieved if function $\sum_{i=0}^{\gamma}[\overline{V}_i]$ with $\overline{V}_i$ as defined in (5.4) is satisfied. In order to fulfill (5.4), then

$$
\begin{aligned}
\sum_{i=0}^{\gamma} \overline{V}_i &= \sum_{i=0}^{\gamma} \left[V_i(\chi_{x+1}) - V_i(\chi_{i,k}) + \alpha_i^{\top}(\chi_{i,k}) R_i(\chi_{i,k}) \alpha_i(\chi_{i,k}) \right] \\
&= \sum_{i=0}^{\gamma} \left[\frac{1}{2} f_i^{\top}(\chi_{i,k}) P_i f_i(\chi_{i,k}) + \frac{2}{2} f_i^{\top}(\chi_{i,k}) P_i g_i(\chi_{i,k}) \alpha_i(\chi_{i,k}) \right. \\
&\quad + \frac{\alpha_i^{\top}(\chi_{i,k}) g_i^{\top}(\chi_{i,k}) P_i g_i(\chi_{i,k}) \alpha_i(\chi_{i,k})}{2} - \frac{x_{i,k}^{\top} P_i \chi_{i,k}}{2} \\
&\quad \left. + \alpha_i^{\top}(\chi_{i,k}) R_i(\chi_{i,k}) \alpha_i(\chi_{i,k}) \right] \\
&= \sum_{i=0}^{\gamma} \left[V_{i,f}(\chi_{i,k}) - \frac{1}{2} P_{i,1}^{\top}(\chi_{i,k}) (R_i(\chi_{i,k}) + P_{i,2}(\chi_{i,k}))^{-1} P_{i,1}(\chi_{i,k}) \right. \\
&\quad \left. + \frac{1}{4} P_{i,1}^{\top}(\chi_{i,k}) (R_i(\chi_{i,k}) + P_{i,2}(\chi_{i,k}))^{-1} P_{i,1}(\chi_{i,k}) \right] \\
&= \sum_{i=0}^{\gamma} \left[V_{i,f}(\chi_{i,k}) - \frac{1}{4} P_{i,1}^{\top}(\chi_{i,k}) (R_i(\chi_{i,k}) + P_{i,2}(\chi_{i,k}))^{-1} P_{i,1}(\chi_{i,k}) \right].
\end{aligned}
\tag{5.14}
$$

Selecting P_i such that $\sum_{i=0}^{\gamma}[\overline{V}_i] \leq 0$, stability of $\sum_{i=0}^{\gamma}[\chi_{i,k}] = 0$ where $\chi_{i,k} = 0$ is guaranteed. Furthermore, by means of P_i, we can achieve a desired negativity amount [5] for the closed-loop function $\overline{V}_i$ in (5.14). This negativity amount can be bounded using a positive definite matrix Q_i as follows:

$$\sum_{i=0}^{\gamma} \overline{V}_i = \sum_{i=0}^{\gamma} \left[V_{i,f}(\chi_{i,k}) - \frac{1}{4} P_{i,1}^{\top}(\chi_{i,k}) (R_i(\chi_{i,k}) + P_{i,2}(\chi_{i,k}))^{-1} P_{i,1}(\chi_{i,k}) \right] \tag{5.15}$$

$$\leq -\sum_{i=0}^{\gamma} x_{i,k}^{\top} Q_i \chi_{i,k} \leq -\sum_{i=0}^{\gamma} \lambda_{\min}(Q_i) \|\chi_{i,k}\|^2$$

$$\leq -\sum_{i=0}^{\gamma} \zeta_{i,Q} \|\chi_{i,k}\|^2, \quad \sum_{i=0}^{\gamma} \zeta_{i,Q} = \sum_{i=0}^{\gamma} \lambda_{\min}(Q_i)$$

Thus, from (5.15) follows condition (5.11). Considering (5.14)–(5.15), if $\sum_{i=0}^{\gamma}[\overline{V}_i]$ $= \sum_{i=0}^{\gamma}[V_i(\chi_{i,k+1}) - V_i(\chi_{i,k}) + \alpha_i^{\top}(\chi_{i,k})R_i(\chi_{i,k})\alpha_i(\chi_{i,k})] \leq -\sum_{i=0}^{\gamma}[\zeta_{i,Q}\|\chi_{i,k}\|^2]$, then $\sum_{i=0}^{\gamma}[\Delta V_i] = \sum_{i=0}^{\gamma}[V_i(\chi_{i,k+1}) - V_i(\chi_{i,k})] \leq \sum_{i=0}^{\gamma}[-\zeta_{i,Q}\|\chi_{i,k}\|^2]$. Moreover, as $\sum_{i=0}^{\gamma}[V_i(\chi_{i,k})]$ are radially unbounded functions, then the solutions $\sum_{i=0}^{\gamma}[\chi_{i,k}] = 0$ of the closed-loop system (5.2) with (5.7) as input, are globally exponentially stable according to Theorem (2.13).

When function $-l_i(\chi_{i,k})$ is set to be the (5.4) right-hand side, i.e., $\sum_{i=0}^{\gamma}[l_i(\chi_{i,k})]$ $= -\sum_{i=0}^{\gamma} \overline{V}_i \,|_{u_{i,k}^* = \alpha_i(\chi_{i,k})} \geq 0$, then $V_i(\chi_{i,k})$ is a solution of the HJB equations (2.31) as established in Definition 5.1.

In order to establish optimality, considering that (5.7) stabilizes (5.2), and substituting $l_i(\chi_{i,k})$ in (5.12), we obtain

$$\sum_{i=0}^{\gamma} \mathscr{J}_i(\chi_{i,k}) = \sum_{i=0}^{\gamma} \sum_{k=0}^{\infty} (l_i(\chi_{i,k}) + u_{i,k}^{\top} R_i(\chi_{i,k})u_{i,k})$$

$$= \sum_{i=0}^{\gamma} \sum_{k=0}^{\infty} (-\overline{V}_i + u_{i,k}^{\top} R_i(\chi_{i,k})u_{i,k}) \quad (5.16)$$

$$= -\sum_{i=0}^{\gamma} \left[\sum_{k=0}^{\infty} \left[V_{i,f}(\chi_{i,k}) - \frac{1}{4} P_{i,1}^{\top}(\chi_{i,k}) \right. \right.$$

$$\times (R_l(\chi_{i,k}) + P_{i,2}(\chi_{i,k}))^{-1} P_{i,1}(\chi_{i,k}) \Bigg]$$

$$\left. + \sum_{k=0}^{\infty} u_{i,k}^{\top} R_i(\chi_{i,k})u_{i,k} \right].$$

Now, factorizing (5.16) and then adding the identity matrix $I_{m_i} \in \mathbb{R}^{m_i \times m_i}$ presented as $I_{m_i} = (R_i(\chi_{i,k}) + P_{i,2}(\chi_{i,k}))(R_i(\chi_{i,k}) + P_{i,2}(\chi_{i,k}))^{-1}$, we obtain

$$\sum_{i=0}^{\gamma} \mathscr{J}_i(\chi_{i,k}) = -\sum_{i=0}^{\gamma} \sum_{k=0}^{\infty} \left[V_{i,f}(\chi_{i,k}) - \frac{1}{2} P_{i,1}^{\top}(\chi_{i,k}) \right.$$

$$\times (R_i(\chi_{i,k}) + P_{i,2}(\chi_{i,k}))^{-1} P_{i,1}(\chi_{i,k})$$

$$+ \frac{1}{4} P_{i,1}^{\top}(\chi_{i,k})(R_i(\chi_{i,k}) + P_{i,2}(\chi_{i,k}))^{-1}$$

$$\times P_{i,2}(\chi_{i,k})(R_i(\chi_{i,k}) + P_{i,2}(\chi_{i,k}))^{-1} P_{i,1}(\chi_{i,k})$$

$$+ \frac{1}{4} P_{i,1}^{\top}(\chi_{i,k})(R_i(\chi_{i,k}) + P_{i,2}(\chi_{i,k}))^{-1}$$

$$\times R_i(\chi_{i,k})(R_i(\chi_{i,k}) + P_{i,2}(\chi_{i,k}))^{-1} P_{i,1}(\chi_{i,k}) \Big]$$

$$+ \sum_{i=0}^{\gamma} \sum_{k=0}^{\infty} u_{i,k}^{\top} R_i(\chi_{i,k}) u_{i,k} \tag{5.17}$$

Being $\alpha_i(\chi_{i,k}) = -\frac{1}{2}(R_i(\chi_{i,k}) + P_{i,2}(\chi_{i,k}))^{-1} P_{i,1}(\chi_{i,k})$, then (5.17) becomes

$$\begin{aligned}
\sum_{i=0}^{\gamma} \mathcal{J}_i(\chi_{i,k}) &= -\sum_{i=0}^{\gamma} \sum_{k=0}^{\infty} \big[V_{i,f}(\chi_{i,k}) + P_{i,1}^{\top}(\chi_{i,k}) \alpha_i(\chi_{i,k}) \\
&\quad + \alpha_i^{\top}(\chi_{i,k}) P_{i,2}(\chi_{i,k}) \alpha_i(\chi_{i,k}) \big] \\
&\quad + \sum_{i=0}^{\gamma} \sum_{k=0}^{\infty} \big[u_{i,k}^{\top} R_i(\chi_{i,k}) u_{i,k} - \alpha_i^{\top}(\chi_{i,k}) R_i(\chi_{i,k}) \alpha_i(\chi_{i,k}) \big] \\
&= -\sum_{i=0}^{\gamma} \bigg[\sum_{k=0}^{\infty} \big[V_i(\chi_{i,k+1}) - V_i(\chi_{i,k}) \big] \tag{5.18} \\
&\quad + \sum_{k=0}^{\infty} \big[u_{i,k}^{\top} R_i(\chi_{i,k}) u_{i,k} - \alpha_i^{\top}(\chi_{i,k}) R_i(\chi_{i,k}) \alpha_i(\chi_{i,k}) \big] \bigg]
\end{aligned}$$

After evaluating which addition for $k = 0$, then (5.18) can be written as

$$\begin{aligned}
\sum_{i=0}^{\gamma} \mathcal{J}_i(\chi_{i,k}) &= -\sum_{i=0}^{\gamma} \bigg[\sum_{k=1}^{\infty} \big[V_i(\chi_{i,k+1}) - V_i(\chi_{i,k}) \big] - V_i(\chi_{i,1}) + V_i(\chi_{i,0}) \bigg] \\
&\quad + \sum_{i=0}^{\gamma} \bigg[\sum_{k=0}^{\infty} \big[u_{i,k}^{\top} R_i(\chi_{i,k}) u_{i,k} - \alpha_i^{\top}(\chi_{i,k}) R_i(\chi_{i,k}) \alpha_i(\chi_{i,k}) \big] \bigg] \\
&= -\sum_{i=0}^{\gamma} \bigg[\sum_{k=2}^{\infty} \big[V_i(\chi_{i,k+1} - V_i(\chi_{i,k})) \big] - V_i(\chi_{i,2}) \\
&\quad + V_i(\chi_{i,1}) - V_i(\chi_{i,1}) + V_i(\chi_{i,0}) \bigg] \tag{5.19} \\
&\quad + \sum_{i=0}^{\gamma} \bigg[\sum_{k=0}^{\infty} \big[u_{i,k}^{\top} R_i(\chi_{i,k}) u_{i,k} - \alpha_i^{\top}(\chi_{i,k}) R_i(\chi_{i,k}) \alpha_i(\chi_{i,k}) \big] \bigg].
\end{aligned}$$

For notation convenience in (5.19), the upper limit ∞ will treated as $N \to \infty$, and thus we obtain

$$\begin{aligned}
\sum_{i=0}^{\gamma} \mathcal{J}_i(\chi_{i,k}) &= \sum_{i=0}^{\gamma} \bigg[-V_i(\chi_{i,N}) + V_i(\chi_{i,N-1}) - V_i(\chi_{i,N-1}) + V_i(\chi_{i,0}) \\
&\quad + \lim_{N \to \infty} \sum_{k=0}^{N} \big[u_{i,k}^{\top} R_i(\chi_{i,k}) u_{i,k} - \alpha_i^{\top}(\chi_{i,k}) R_i(\chi_{i,k}) \alpha_i(\chi_{i,k}) \big] \bigg]
\end{aligned}$$

$$= \sum_{i=0}^{\gamma} \left[-V_i(\chi_{i,N}) + V_i(\chi_{i,0}) \right. \tag{5.20}$$

$$\left. + \lim_{N \to \infty} \sum_{k=0}^{N} \left[u_{i,k}^\top R_i(\chi_{i,k}) u_{i,k} - \alpha_i^\top(\chi_{i,k}) R_i(\chi_{i,k}) \alpha_i(\chi_{i,k}) \right] \right].$$

Let consider $N \to \infty$ and note that $V_i(\chi_{i,N}) \to 0$ for all $\chi_{i,0}$, then

$$\sum_{i=0}^{\gamma} \mathscr{J}_i(\chi_{i,k}) = \sum_{i=0}^{\gamma} \left[V_i(\chi_{i,0}) + \sum_{k=0}^{\infty} \left[u_{i,k}^\top R_i(\chi_{i,k}) u_{i,k} - \alpha_i^\top(\chi_{i,k}) R_i(\chi_{i,k}) \alpha_i(\chi_{i,k}) \right] \right] \tag{5.21}$$

Thus, the minima values of (5.21) are reached with $\sum_{i=0}^{\gamma}[u_{i,k}] = \sum_{i=0}^{\gamma}[\alpha_i(\chi_{i,k})]$. Hence, the control laws (5.7) minimizes the cost function (5.12). The optimal value function of (5.12) is $\sum_{i=0}^{\gamma}[\mathscr{J}_i^*(\chi_{i,0})] = \sum_{i=0}^{\gamma}[V_i(\chi_{i,0})]$ for all $\chi_{i,0}$.   ∎

Optimal control will be in general of the form (5.3) and the minimum value of the performance index will be function of the initial state $\chi_{i,0}$. If system (5.2) and the control law (5.3) are given, we shall say that the pair $\{V_i(\chi_{i,k}), l_i(\chi_{i,k})\}$ is a solution to the *inverse optimal control problem* if the performance index (2.24) is minimized by (5.3), with the minimum value $V_i(\chi_{i,0})$.

5.2   Neural Network Identifier

Using the structure of system (5.1), we propose the following neural network model:

$$x_{i,k+1}^j = W_{i,k}^j z_i(\chi_k, u_k) + W_i^{\prime j} \psi_i^j(\chi_{i,k}^{j+1}, u_{i,k}) \tag{5.22}$$

where $x_i = [x_i^{1\top}, x_i^{2\top}, \ldots, x_i^{n_i\top}]$ is the i-th block neuron state with the same properties that (5.1), $W_{i,k}^j$ are the adjustable weight matrices, $W_{i,k}^{\prime j}$ are matrices with fixed parameters and $rank(W_{i,k}^{\prime j}) = n_{ij}$, with $j = 1, \ldots, n_i$, $i = 1, \ldots, \gamma$; ψ denotes a linear function of x or u corresponding to the plant structure (5.1) or external inputs to the network, respectively.

It is worth to note that, (5.22) constitutes a series-parallel identifier [3, 8] and does not consider explicitly the interconnection terms, whose effects are compensated by the neural network weights update.

Proposition 5.4 *The tracking of a desired trajectory $x_{i\delta}^j$, defined in terms of the plant state χ_i^j formulated as (5.1) can be established as the following inequality [4]*

$$\| x_{i\delta}^j - \chi_i^j \| \leq \| x_i^j - \chi_i^j \| + \| x_{i\delta}^j - x_i^j \| \tag{5.23}$$

where $\| \bullet \|$ stands for the Euclidean norm, $i = 1, \ldots, \gamma$, $j = 1, \ldots, n_i$; $x_{i\delta}^j - \chi_i^j$ is the system output tracking error; $x_i^j - \chi_i^j$ is the output identification error; and $x_{i\delta}^j - x_i^j$ is the RHONN output tracking error.

We establish the following requirements for tracking solution:

Requirement 5.1

$$\lim_{t \to \infty} \| x_i^j - \chi_i^j \| \leq \zeta_i^j \tag{5.24}$$

with ζ_i^j a small positive constant.

Requirement 5.2

$$\lim_{t \to \infty} \| x_{i\delta}^j - x_i^j \| = 0. \tag{5.25}$$

An on-line decentralized neural identifier based on (5.22) ensures (5.24), while (5.25) is guaranteed by a discrete-time decentralized inverse optimal control.

It is possible to establish Proposition 5.4 due to separation principle for discrete-time nonlinear systems [9]. It is clear that the decentralized inverse optimal control for stabilization is a particular case of trajectory tracking when $x_{i\delta}$ is matched to an equilibrium point.

5.3   On-Line Learning Law

We use an EKF-based training algorithm described by [1, 2, 6]

$$K_{iq,k}^j = P_{iq,k}^j H_{iq,k}^j M_{iq,k}^j$$
$$w_{iq,k+1}^j = w_{iq,k}^j + \eta_{iq}^j K_{iq,k}^j e_{iq,k}^j \tag{5.26}$$
$$P_{iq,k+1}^j = P_{iq,k}^j - K_{iq,k}^j H_{iq,k}^{j\top} P_{iq,k}^j + Q_{iq,k}^j$$

with

$$M_{iq,k}^j = \left[R_{iq,k}^j + H_{iq,k}^{j\top} P_{iq,k}^j H_{iq,k}^j \right]^{-1} \tag{5.27}$$

$$e_{iq,k}^j = \chi_{iq,k}^j - x_{iq,k}^j \tag{5.28}$$

where $e_{iq,k}^j$ is the identification error, $P_{iq,k+1}^j$ is the state estimation prediction error covariance matrix, $w_{iq,k}^j$ is the jq-th weight (state) of the i-th subsystem, η_{iq}^j is a design parameter such that $0 \leq \eta_{iq}^j \leq 1$, $\chi_{iq,k}^j$ is the jq-th plant state, $x_{iq,k}^j$ is the jq-th neural network state, q is the number of states, $K_{iq,k}^j$ is the Kalman gain matrix, $Q_{iq,k}^j$ is the measurement noise covariance matrix, $R_{iq,k}^j$ is the state noise covariance matrix, and $H_{iq,k}^j$ is a matrix, in which each entry of $(H_{q,k}^j)$ is the derivative of jq-th neural network state $(x_{iq,k}^j)$, with respect to all adjustable weights $(w_{iq,k}^j)$, as follows

$$H_{q,k}^j = \left[\frac{\partial x_{iq,k}^j}{\partial w_{iq,k}^j} \right]_{w_{iq,k}^j = w_{iq,k+1}^j}^{\top}, \qquad (5.29)$$

$$i = 1, \dots, \gamma \text{ and } j = 1, \dots, n_i$$

Usually P_{iq}^j, Q_{iq}^j and R_{iq}^j are initialized as diagonal matrices, with entries $P_{iq}^j(0)$, $Q_{iq}^j(0)$ and $R_{iq}^j(0)$, respectively [7]. It is important to note that $H_{iq,k}^j$, $K_{iq,k}^j$ and $P_{iq,k}^j$ for the EKF are bounded [11]. Then the dynamics of (5.28) can be expressed as

$$e_{iq,k}^j = \widetilde{w}_{iq,k}^j \varphi_{iq,k}^j (x_k, u_k) + \epsilon_{\varphi_{iq,k}^j} \qquad (5.30)$$

on the other hand, the dynamics of weight estimation error $\widetilde{w}_{iq,k}^j$ is

$$\widetilde{w}_{iq,k+1}^j = \widetilde{w}_{iq,k}^j - \eta_{iq}^j K_{iq,k}^j e_{iq,k}^j \qquad (5.31)$$

For the case when i is fixed, the stability analysis for the i-th subsystem of RHONN (5.22) to identify the i-th subsystem of nonlinear plant (5.2) in absence of interconnections, is based on the Theorem 3.2.

Let consider the RHONN (5.22) which identify the nonlinear plant (5.1) in presence of interconnections, is based on the Theorem 3.3.

References

1. Alanis, A.Y., Sanchez, E.N., Loukianov, A.G.: Real-time output tracking for induction motors by recurrent high-order neural network control. In: Proceedings of 17th Mediterranean Conference on Control and Automation (MED 2009), pp. 868–873. Thessaloniki, Greece (2009)
2. Feldkamp, L.A., Prokhorov, D.V., Feldkamp, T.M.: Simple and conditioned adaptive behavior from Kalman filter trained recurrent networks. Neural Netw. 16(5), 683–689 (2003)
3. Felix, R.A.: Variable Structure Neural Control. PhD thesis, Cinvestav, Unidad Guadalajara, Guadalajara, Jalisco, Mexico (2003)
4. Felix, R.A., Sanchez, E.N., Loukianov, A.G.: Avoiding controller singularities in adaptive recurrent neural control. Proceedings of the 16th IFAC World Congress, pp. 1095–1100. Czech Republic, Prague (2005)
5. Freeman, R.A., Primbs, J.A.: Control Lyapunov functions: new ideas from an old source. Proceedings of the 35th IEEE Conference on Decision and Control, pp. 3926–3931. Portland, OR, USA (1996)
6. Grover, R., Hwang, P.Y.C.: Introduction to Random Signals and Applied Kalman Filtering. Wiley Inc, New York (1992)
7. Haykin, S.: Kalman Filtering and Neural Networks. Wiley Inc, New York (2001)
8. Ioannou, P., Sun, J.: Robust Adaptive Control. Prentice Hall Inc, Upper Saddle River (1996)

9. Lin, W., Byrnes, C.I.: Design of discrete-time nonlinear control systems via smooth feedback. IEEE Trans. Autom. Control **39**(11), 2340–2346 (1994)
10. Sanchez, E.N., Ornelas-Tellez, F.: Discrete-Time Inverse Optimal Control for Nonlinear Systems. CRC Press, Boca Raton (2013)
11. Song, Y., Grizzle, J.W.: The extended Kalman filter as a local asymptotic observer for discrete-time nonlinear systems. J. Math. Syst. Estim. Control **5**(1), 59–78 (1995)

Chapter 6
Decentralized Inverse Optimal Control for Trajectory Tracking

6.1 Trajectory Tracking Optimal Control

For system (2.23), it is desired to determine a control law $u_k = \bar{u}(x_k)$ which minimizes a cost functional. The following cost functional is associated with trajectory tracking for system (2.23):

$$\mathscr{J}(z_k) = \sum_{n=k}^{\infty}(l(z_n) + u_n^\top R u_n) \tag{6.1}$$

where $z_k = \chi_k - \chi_{\delta,k}$ with $\chi_{\delta,k}$ as the desired trajectory for χ_k; $z_k \in \mathfrak{R}^n$; $\mathscr{J}(z_k) :$ $\mathfrak{R}^n \to \mathfrak{R}^+$; $l(z_k) : \mathfrak{R}^n \to \mathfrak{R}^+$ is a positive semidefinite function and $R : \mathfrak{R}^n \to \mathfrak{R}^{m \times m}$ is a real symmetric positive definite weighting matrix. The meaningful cost functional (6.1) is a performance measure [6]. The entries of R may be functions of the system state in order to vary the weighting on control efforts according to the state value [6]. Considering the state feedback control approach, we assume that the full state χ_k is available.

Equation (6.1) can be rewritten as

$$\mathscr{J}(z_k) = l(z_k) + u_k^\top R u_k + \sum_{n=k+1}^{\infty} l(z_n) + u_n^\top R u_n \tag{6.2}$$

$$= l(z_k) + u_k^\top R u_k + \mathscr{J}(z_{k+1})$$

where we require the boundary condition $\mathscr{J}(0) = 0$ so that $\mathscr{J}(z_k)$ becomes a Lyapunov function [1, 10]. The value of $\mathscr{J}(z_k)$, if finite, then it is a function of the initial state z_0. When $\mathscr{J}(z_k)$ is at its minimum, which is denoted as $\mathscr{J}^*(z_k)$, it is named the optimal value function, and it will be used as a Lyapunov function, i.e., $\mathscr{J}(z_k) \triangleq V(z_k)$.

From Bellman's optimality principle [2, 7], it is known that, for the infinite horizon optimization case, the value function $V(z_k)$ becomes time invariant and satisfies the

© Springer International Publishing Switzerland 2017
R. Garcia-Hernandez et al., *Decentralized Neural Control: Application to Robotics*,
Studies in Systems, Decision and Control 96, DOI 10.1007/978-3-319-53312-4_6

discrete-time (DT) Bellman equation [1, 2, 8]

$$V(z_k) = \min_{u_k} \{l(z_k) + u_k^\top R u_k + V(z_{k+1})\} \tag{6.3}$$

where $V(z_{k+1})$ depends on both z_k and u_k by means of z_{k+1} in (2.23). Note that the DT Bellman equation is solved backward in time [1]. In order to establish the conditions that the optimal control law must satisfy, we define the discrete-time Hamiltonian $\mathcal{H}(z_k, u_k)$ [5] as

$$\mathcal{H}(z_k, u_k) = l(z_k) + u_k^\top R u_k + V(z_{k+1}) - V(z_k). \tag{6.4}$$

The Hamiltonian is a method to adjoin constraint (2.23) to the performance index (6.1), and then, solving the optimal control problem by minimizing the Hamiltonian without constraints [7].

A necessary condition that the optimal control law u_k should satisfy is $\frac{\partial \mathcal{H}(z_k, u_k)}{\partial u_k}$ $= 0$ [6], which is equivalent to calculate the gradient of (6.3) right-hand side with respect to u_k, then

$$
\begin{aligned}
0 &= 2Ru_k + \frac{\partial V(z_{k+1})}{\partial u_k} \\
&= 2Ru_k + \frac{\partial z_{k+1}}{\partial u_k} \frac{\partial V(z_{k+1})}{\partial z_{k+1}} \\
&= 2Ru_k + g^\top(\chi_k) \frac{\partial V(z_{k+1})}{\partial z_{k+1}}
\end{aligned}
\tag{6.5}
$$

Therefore, the optimal control law is formulated as

$$u_k^* = -\frac{1}{2} R^{-1} g^\top(\chi_k) \frac{\partial V(z_{k+1})}{\partial z_{k+1}} \tag{6.6}$$

with the boundary condition $V(0) = 0$; u_k^* is used when we want to emphasize that u_k is optimal. Moreover, if $\mathcal{H}(z_k, u_k)$ has a quadratic form in u_k and $R > 0$, then

$$\frac{\partial^2 \mathcal{H}(z_k, u_k)}{\partial u_k^2} > 0$$

holds as a sufficient condition such that optimal control law (6.6) (globally [6]) minimizes $\mathcal{H}(z_k, u_k)$ and the performance index (6.1) [7].

Substituting (6.6) into (6.3), we obtain the discrete-time Hamilton–Jacobi–Bellman (HJB) equation described by

$$
\begin{aligned}
V(z_k) &= l(z_k) + V(z_{k+1}) \\
&+ \frac{1}{4} \frac{\partial V^\top(z_{k+1})}{\partial z_{k+1}} g(\chi_k) R^{-1} g^\top(\chi_k) \frac{\partial V(z_{k+1})}{\partial z_{k+1}}
\end{aligned}
\tag{6.7}
$$

which can be rewritten as

$$0 = l(z_k) + V(z_{k+1}) - V(z_k) \qquad (6.8)$$
$$+ \frac{1}{4} \frac{\partial V^\top(z_{k+1})}{\partial z_{k+1}} g(\chi_k) R^{-1} g^\top(\chi_k) \frac{\partial V(z_{k+1})}{\partial z_{k+1}}$$

Solving the HJB partial-differential equation (6.8) is not straightforward; this is one of the main disadvantages of discrete-time optimal control for nonlinear systems. To overcome this problem, we propose the inverse optimal control.

6.2 Decentralized Trajectory Tracking Optimal Control

Let consider a class of disturbed discrete-time nonlinear and interconnected system

$$\chi_{i,k+1}^j = f_i^j(\chi_{i,k}^j) + g_i^j(\chi_{i,k}^j) u_{i,k} + \Gamma_{i\ell,k}^j(\chi_\ell) \qquad (6.9)$$

where $i = 1, \ldots, \gamma$, $j = 1, \ldots, n_i$, $\chi_i \in \Re^{n_i}$, $\chi_i = [\chi_i^{1\top} \chi_i^{2\top} \cdots \chi_i^{r\top}]^\top$, $\chi_i^j \in \Re^{n_{ij} \times 1}$ $u_i \in \Re^{m_i}$; γ is the number of subsystems, χ_ℓ reflect the interaction between the i-th and the ℓ-th subsystem with $1 \le \ell \le \gamma$. We assume that f_i, B_i and Γ_i are smooth and bounded functions, $f_i^j(0) = 0$ and $B_i^j(0) = 0$. Without loss of generality, $\chi = 0$ is an equilibrium point of (6.9), which is of be used later.

For the inverse optimal control approach, let consider the discrete-time affine in the input nonlinear system:

$$\chi_{i,k+1} = f_i(\chi_{i,k}) + g_i(\chi_{i,k}) u_{i,k}, \quad \chi_{i,0} = \chi_i(0) \qquad (6.10)$$

with $i = 1, \ldots, \gamma$; γ is the number of subsystems. Where $\chi_{i,k} \in \Re^{n_i}$ are the states of the systems, $u_{i,k} \in \Re^{m_i}$ are the control inputs, $f_i(\chi_{i,k}) : \Re^{n_i} \to \Re^{n_i}$ and $g(\chi_k) : \Re^{n_i} \to \Re^{n_i \times m_i}$ are smooth maps, the subscript $k \in \mathbb{Z}^+ \cup 0 = \{0, 1, 2, \ldots\}$ will stand for the value of the functions and/or variables at the time k. We establish the following assumptions and definitions which allow the inverse optimal control solution via the CLF approach.

Assumption 6.1 The full state of system (6.10) is measurable.

Definition 6.1 ([9]) Consider the tracking error $z_{i,k} = \chi_{i,k} - \chi_{i\delta,k}$ with $\chi_{i\delta,k}$ as the desired trajectory for $\chi_{i,k}$. Let define the control law

$$u_{i,k}^* = -\frac{1}{2} R_i^{-1} g_i^\top(\chi_{i,k}) \frac{\partial V_i(z_{i,k+1})}{\partial z_{i,k+1}} \qquad (6.11)$$

which will be inverse optimal stabilizing along the desired trajectory $\chi_{i\delta,k}$ if:

- (i) it achieves (global) asymptotic stability for system (6.10) along reference $\chi_{i\delta,k}$;

- (ii) $V_i(z_{i,k})$ is (radially unbounded) positive definite function such that inequality

$$\overline{V}_i := V_i(z_{i,k+1}) - V_i(z_{i,k}) + u_{i,k}^{*\top} R_i u_{i,k}^* \leq 0 \qquad (6.12)$$

is satisfied.

When $l_i(z_{i,k}) := -\overline{V}_i \leq 0$ is selected, then $V_i(z_{i,k})$ is a solution for the HJB equation

$$
\begin{aligned}
0 = \; & l_i(z_{i,k}) + V_i(z_{i,k+1}) - V_i(z_{i,k}) \qquad (6.13) \\
& + \frac{1}{4} \frac{\partial V_i^\top(z_{i,k+1})}{\partial z_{i,k+1}} g_i(\chi_{i,k}) R^{-1} g_i^\top(\chi_{i,k}) \frac{\partial V_i(z_{i,k+1})}{\partial z_{i,k+1}}
\end{aligned}
$$

and the cost functional (6.1) is minimized.

As established in Definition 6.1, the inverse optimal control law for trajectory tracking is based on the knowledge of $V_i(z_{i,k})$. Thus, we propose a CLF $V_i(z_{i,k})$, such that (i) and (ii) are guaranteed. That is, instead of solving (6.8) for $V_i(z_{i,k})$, a quadratic CLF candidate $V_i(z_{i,k})$ is proposed with the form:

$$V_i(z_{i,k}) = \frac{1}{2} z_{i,k}^\top P_i z_{i,k}, \quad P_i = P_i^\top > 0 \qquad (6.14)$$

for control law (6.11) in order to ensure stability of the tracking error $z_{i,k}$, where

$$
\begin{aligned}
z_{i,k} &= \chi_{i,k} - \chi_{i\delta,k} \qquad (6.15) \\
&= \begin{bmatrix} (\chi_{i,k}^1 - \chi_{i\delta,k}^1) \\ (\chi_{i,k}^2 - \chi_{i\delta,k}^2) \\ \vdots \\ (\chi_{i,k}^n - \chi_{i\delta,k}^n) \end{bmatrix}
\end{aligned}
$$

Moreover, it will be established that control law (6.11) with (6.14), which is referred to as the inverse optimal control law, optimizes a cost functional of the form (6.1).

Consequently, by considering $V_i(\chi_{i,k})$ as in (6.14), control law (6.11) takes the following form:

$$
\begin{aligned}
\alpha_i(\chi_{i,k}) := u_{i,k}^* &= -\frac{1}{4} R_i g_i^\top(\chi_{i,k}) \frac{\partial z_{i,k+1}^\top P_i z_{i,k+1}}{\partial z_{i,k+1}} \\
&= -\frac{1}{2} R_i g_i^\top(\chi_{i,k}) P_i z_{i,k+1} \qquad (6.16) \\
&= -\frac{1}{2} (R_i(\chi_{i,k}) + \frac{1}{2} g_i^\top(\chi_{i,k}) P_i g_i(\chi_{i,k}))^{-1} \\
& \quad \times g_i^\top(\chi_{i,k}) P_i (f_i(\chi_{i,k}) - \chi_{i\delta,k+1})
\end{aligned}
$$

It is worth to point out that P_i and R_i are positive definite and symmetric matrices; thus, the existence of the inverse in (6.16) is ensured.

Once we have proposed a CLF for solving the inverse optimal control in accordance with Definition 6.1, the respective solution is presented, for which P_i is considered a fixed matrix.

Lemma 6.2 *Consider the affine discrete-time nonlinear system (6.10) with $i = 1$, If there exists a matrix $P_i = P_i^\top > 0$ such that the following inequality holds:*

$$
\begin{aligned}
&\frac{1}{2} f_i^\top(\chi_{i,k}) P_i f_i(\chi_{i,k}) + \frac{1}{2} \chi_{i\delta,k+1}^\top P_i \chi_{i\delta,k+1} - \chi_{i,k}^\top P_i \chi_{i\delta,k}^\top \\
&- \frac{1}{2} \chi_{i\delta,k}^\top P_i \chi_{i\delta,k} - \frac{1}{4} P_{i1}^\top(\chi_{i,k}, \chi_{i\delta,k})(R_i + P_{i2}(\chi_{i,k}))^{-1} \\
&\times P_{i1}(\chi_{i,k}, \chi_{i\delta,k}) \\
&\leq -\frac{1}{2} \|P_i\| \|f_i(\chi_{i,k})\|^2 - \frac{1}{2} \|P_i\| \|\chi_{i\delta,k+1}\|^2 \\
&\quad - \frac{1}{2} \|P_i\| \|\chi_{i,k}\|^2 - \frac{1}{2} \|P_i\| \|\chi_{i\delta,k}\|^2
\end{aligned}
\tag{6.17}
$$

where $P_{i,1}(\chi_{i,k}, \chi_{i\delta,k})$ and $P_{i,2}(\chi_{i,k})$ are defined as

$$
P_{i,1}(\chi_{i,k}, \chi_{i\delta,k}) = g_i^\top(\chi_{i,k}) P_i (f_i(\chi_{i,k}) - \chi_{i\delta,k+1})
\tag{6.18}
$$

and

$$
P_{i,2}(\chi_{i,k}) = \frac{1}{2} g_i^\top(\chi_{i,k}) P_i g_i(\chi_{i,k})
\tag{6.19}
$$

respectively, then system (6.10) with control law (6.16) guarantees asymptotic trajectory tracking along the desired trajectory $\chi_{i\delta,k}$, where $z_{i,k+1} = \chi_{i,k+1} - \chi_{i\delta,k+1}$.

Moreover, with (6.14) as a CLF, this control law is inverse optimal in the sense that it minimizes the meaningful functional given by

$$
\mathcal{J}_i(z_{i,k}) = \sum_{k=0}^{\infty} (l_i(z_{i,k}) + u_{i,k}^\top R_i(z_{i,k}) u_{i,k})
\tag{6.20}
$$

with

$$
l_i(z_{i,k}) = -\overline{V}_i|_{u_{i,k}^* = \alpha_i(z_{i,k})}
\tag{6.21}
$$

and optimal value function $\mathcal{J}_i(z_{i,k}) = V_i(z_0)$.

This lemma is adapted from [9] for each isolated subsystem, which allows to establish the following theorem.

Theorem 6.3 *Consider the affine discrete-time nonlinear system (6.10) with $i = 1, \ldots, \gamma$; γ is the number of subsystems. If there exists matrices $P_i = P_i^\top > 0$ such that the following inequality holds:*

$$\sum_{i=0}^{\gamma} \left[\frac{1}{2} f_i^\top(\chi_{i,k}) P_i f_i(\chi_{i,k}) + \frac{1}{2} \chi_{i\delta,k+1}^\top P_i \chi_{i\delta,k+1} \right.$$

$$-\chi_{i,k}^\top P_i \chi_{i\delta,k} - \frac{1}{2} \chi_{i\delta,k}^\top P_i \chi_{i\delta,k}$$

$$\left. -\frac{1}{4} P_{i1}^\top(\chi_{i,k}, \chi_{i\delta,k})(R_i + P_{i2}(\chi_{i,k}))^{-1} P_{i1}(\chi_{i,k}, \chi_{i\delta,k}) \right]$$

$$\leq \sum_{i=0}^{\gamma} \left[-\frac{1}{2} \|P_i\| \|f_i(\chi_{i,k})\|^2 - \frac{1}{2} \|P_i\| \|\chi_{i\delta,k+1}\|^2 \right.$$

$$\left. -\frac{1}{2} \|P_i\| \|\chi_{i,k}\|^2 - \frac{1}{2} \|P_i\| \|\chi_{i\delta,k}\|^2 \right] \tag{6.22}$$

where $\sum_{i=0}^{\gamma}[P_{i,1}(\chi_{i,k}, \chi_{i\delta,k})]$ and $\sum_{i=0}^{\gamma}[P_{i,2}(\chi_{i,k})]$ are defined as

$$\sum_{i=0}^{\gamma} \left[P_{i,1}(\chi_{i,k}, \chi_{i\delta,k}) \right] = \sum_{i=0}^{\gamma} \left[g_i^\top(\chi_{i,k}) P_i (f_i(\chi_{i,k}) - \chi_{i\delta,k+1}) \right] \tag{6.23}$$

and

$$\sum_{i=0}^{\gamma} \left[P_{i,2}(\chi_{i,k}) \right] = \sum_{i=0}^{\gamma} \left[\frac{1}{2} g_i^\top(\chi_{i,k}) P_i g_i(\chi_{i,k}) \right] \tag{6.24}$$

respectively, then system (6.10) with control law (6.16) guarantees asymptotic trajectory tracking along the desired trajectory $\chi_{i\delta,k}$, where $z_{i,k+1} = \chi_{i,k+1} - \chi_{i\delta,k+1}$.

Moreover, with (6.14) as a CLF, this control law is inverse optimal in the sense that it minimizes the meaningful functional given by

$$\sum_{i=0}^{\gamma} \mathscr{J}_i(z_{i,k}) = \sum_{i=0}^{\gamma} \sum_{k=0}^{\infty} (l_i(z_{i,k}) + u_{i,k}^\top R_i(z_{i,k}) u_{i,k}) \tag{6.25}$$

with

$$\sum_{i=0}^{\gamma} l_i(z_{i,k}) = -\sum_{i=0}^{\gamma} \overline{V}_i|_{u_{i,k}^* = \alpha_i(z_{i,k})} \tag{6.26}$$

and optimal value function $\sum_{i=0}^{\gamma}[\mathscr{J}_i(z_{i,k})] = \sum_{i=0}^{\gamma}[V_i(z_0)]$.

Proof System (6.10) with control law (6.16) and (6.14), must satisfy inequality (6.12). Considering Lemma 6.2 and one step ahead for z_k, it is possible to write

(6.12) as

$$
\sum_{i=0}^{\gamma}[\overline{V}] = \sum_{i=0}^{\gamma} \left[\frac{1}{2} z_{i,k+1}^{\top} P_i z_{i,k+1} - \frac{1}{2} z_{i,k}^{\top} P_i z_{i,k} + u_{i,k}^{*\top} R_i u_{i,k}^{*} \right]
$$

$$
= \sum_{i=0}^{\gamma} \left[\frac{1}{2} (\chi_{i,k+1} - \chi_{i\delta,k+1})^{\top} P_i (\chi_{i,k+1} - \chi_{i\delta,k+1}) \right.
$$

$$
\left. - \frac{1}{2} (\chi_{i,k} - \chi_{i\delta,k})^{\top} P_i (\chi_{i,k} - \chi_{i\delta,k}) + u_{i,k}^{*\top} R_i u_{i,k}^{*} \right] \tag{6.27}
$$

Substituting (2.23) and (6.16) in (6.27)

$$
\sum_{i=0}^{\gamma}[\overline{V}] = \sum_{i=0}^{\gamma} \left[\frac{1}{2} (f_i(\chi_{i,k}) + g_i(\chi_{i,k})u_{i,k}^{*} - \chi_{i\delta,k+1})^{\top} \right.
$$

$$
\times P_i (f_i(\chi_{i,k}) + g_i(\chi_{i,k})u_{i,k}^{*} - \chi_{i\delta,k+1})
$$

$$
\left. - \frac{1}{2} (\chi_{i,k} - \chi_{i\delta,k})^{\top} P_i (\chi_{i,k} - \chi_{i\delta,k}) + u_{i,k}^{*\top} R_i u_{i,k}^{*} \right]
$$

$$
= \sum_{i=0}^{\gamma} \left[\frac{1}{2} f_i^{\top}(\chi_{i,k}) P_i f_i(\chi_{i,k}) + \frac{1}{2} f_i^{\top}(\chi_{i,k}) P_i g_i(\chi_{i,k}) u_{i,k}^{*} \right.
$$

$$
+ \frac{1}{2} u_{i,k}^{*\top} g_i^{\top}(\chi_{i,k}) P_i g_i(\chi_{i,k}) u_{i,k}^{*} + \frac{1}{2} \chi_{i\delta,k+1}^{\top} P_i \chi_{i\delta,k+1}
$$

$$
+ \frac{1}{2} u_{i,k}^{*\top} g_i^{\top}(\chi_{i,k}) P_i f_i(\chi_{i,k}) - \frac{1}{2} f_i^{\top}(\chi_{i,k}) P_i \chi_{i\delta,k+1}
$$

$$
- \frac{1}{2} \chi_{i\delta,k+1} P_i f_i^{\top}(\chi_{i,k}) - \frac{1}{2} u_{i,k}^{*\top} g_i^{\top}(\chi_{i,k}) P_i \chi_{i\delta,k+1}
$$

$$
- \frac{1}{2} \chi_{i\delta,k+1}^{\top} P_i g_i(\chi_{i,k}) u_{i,k}^{*} - \frac{1}{2} \chi_{i,k}^{\top} P_i \chi_{i,k} + \frac{1}{2} \chi_{i\delta,k}^{\top} P_i \chi_{i,k}
$$

$$
\left. - \frac{1}{2} \chi_{i\delta,k}^{\top} P_i \chi_{i\delta,k} + \frac{1}{2} \chi_{i,k}^{\top} P_i \chi_{i\delta,k} + u_{i,k}^{*\top} R_i u_{i,k}^{*} \right] \tag{6.28}
$$

by simplifying, (6.28) becomes

$$
\sum_{i=0}^{\gamma}[\overline{V}] = \sum_{i=0}^{\gamma} \left[\frac{1}{2} f_i^{\top}(\chi_{i,k}) P_i f_i(\chi_{i,k}) + \frac{1}{2} \chi_{i\delta,k+1}^{\top} P_i \chi_{i\delta,k+1} \right.
$$

$$
+ \frac{1}{2} u_{i,k}^{*\top} g_i^{\top}(\chi_{i,k}) P_i g_i(\chi_{i,k}) u_{i,k}^{*} - f_i^{\top}(\chi_{i,k}) P_i \chi_{i\delta,k+1}
$$

$$
+ f_i^{\top}(\chi_{i,k}) P_i g_i(\chi_{i,k}) u_{i,k}^{*} - \chi_{i\delta,k+1}^{\top} P_i g_i(\chi_{i,k}) u_{i,k}^{*}
$$

$$
\left. - \frac{1}{2} \chi_{i,k}^{\top} P_i \chi_{i,k} - \frac{1}{2} \chi_{i\delta,k}^{\top} P_i \chi_{i\delta,k} + \chi_{i,k}^{\top} P_i \chi_{i\delta,k} \right.
$$

$$+u_{i,k}^{*\top} R_i u_{i,k}^* \Bigg]$$

$$= \sum_{i=0}^{\gamma} \Bigg[\frac{1}{2} f_i^{\top}(\chi_{i,k}) P_i f_i(\chi_{i,k}) + \frac{1}{2} \chi_{i\delta,k+1}^{\top} P_i \chi_{i\delta,k+1}$$

$$- f_i^{\top}(\chi_{i,k}) P_i \chi_{i\delta,k+1} + \chi_{i,k}^{\top} P_i \chi_{i\delta,k} - \frac{1}{2} \chi_{i,k}^{\top} P_i \chi_{i,k}$$

$$- \frac{1}{2} \chi_{i\delta,k}^{\top} P_i \chi_{i\delta,k} + P_{i,1}^{\top}(\chi_{i,k}, \chi_{i\delta,k}) u_{i,k}^*$$

$$+ u_{i,k}^{*\top} P_{i,2} u_{i,k}^* + u_{i,k}^{*\top} R_i u_{i,k}^* \Bigg] \tag{6.29}$$

which after using the control law (6.16), (6.29) results in

$$\sum_{i=0}^{\gamma} [\overline{V}] = \sum_{i=0}^{\gamma} \Bigg[\frac{1}{2} f_i^{\top}(\chi_{i,k}) P_i f_i(\chi_{i,k}) + \frac{1}{2} \chi_{i\delta,k+1}^{\top} P_i \chi_{i\delta,k+1}$$

$$- \frac{1}{2} \chi_{i,k}^{\top} P_i \chi_{i,k} - \frac{1}{2} \chi_{i\delta,k}^{\top} P_i \chi_{i\delta,k}$$

$$- \frac{1}{4} P_{i,1}^{\top}(\chi_{i,k}, \chi_{i\delta,k})(R + P_{i,2}(\chi_{i,k}))^{-1} P_{i,1}(\chi_{i,k}, \chi_{i\delta,k})$$

$$- f_i^{\top}(\chi_{i,k}) P_i \chi_{i\delta,k+1} + \chi_{i,k}^{\top} P_i \chi_{i\delta,k} \Bigg] \tag{6.30}$$

Analyzing the sixth and seventh right-hand side (RHS) terms of (6.30) by using the inequality $X^{\top} Y + Y^{\top} X \leq X^{\top} \Lambda X + Y^{\top} \Lambda^{-1} Y$ proved in [9], which is valid for any CLF-Based Inverse Optimal Control for a Class of Nonlinear Systems vector $X \in \Re^{n_i \times 1}$, then for the sixth RHS term of (6.30), we have

$$\sum_{i=0}^{\gamma} \left[f_i^{\top}(\chi_{i,k}) P_i \chi_{i\delta,k+1} \right] \leq$$

$$\sum_{i=0}^{\gamma} \Bigg[\frac{1}{2} \big[f_i^{\top}(\chi_{i,k}) P_i f_i(\chi_{i,k})$$

$$+ (P_i \chi_{i\delta,k+1})^{\top} P_i^{-1} (P_i \chi_{i,\delta,k+1}) \big] \Bigg]$$

$$\leq \sum_{i=0}^{\gamma} \Bigg[\frac{1}{2} \big[f_i^{\top}(\chi_{i,k}) P_i f_i(\chi_{i,k}) + \chi_{i,\delta,k+1}^{\top} P_i \chi_{i,\delta,k+1} \big] \Bigg]$$

$$\leq \sum_{i=0}^{\gamma} \Bigg[\frac{1}{2} \| P_i \| \| f_i(\chi_{i,k}) \|^2 + \frac{1}{2} \| P_i \| \| \chi_{i\delta,k+1} \|^2 \Bigg]. \tag{6.31}$$

The seventh RHS term of (6.30) becomes

$$\sum_{i=0}^{\gamma} \left[\chi_{i,k}^{\top} P_i \chi_{i\delta,k} \right] \leq$$

$$\sum_{i=0}^{\gamma} \left[\frac{1}{2} \left[\chi_{i,k}^{\top} P_i \chi_{i,k} + (P_i \chi_{i\delta,k})^{\top} P_i^{-1} (P_i \chi_{i,\delta,k}) \right] \right]$$

$$\leq \sum_{i=0}^{\gamma} \left[\frac{1}{2} \left[\chi_{i,k}^{\top} P_i \chi_{i,k} + \chi_{i\delta,k}^{\top} P_i \chi_{i,\delta,k} \right] \right] \tag{6.32}$$

From (6.32), the following expression holds

$$\sum_{i=0}^{\gamma} \left[\frac{1}{2} \left[\chi_{i,k}^{\top} P_i \chi_{i,k} + \chi_{i\delta,k}^{\top} P_i \chi_{i,\delta,k} \right] \right] \leq$$

$$\sum_{i=0}^{\gamma} \left[\frac{1}{2} \| P_i \| \| \chi_{i,k} \|^2 + \frac{1}{2} \| P_i \| \| \chi_{i\delta,k} \|^2 \right] \tag{6.33}$$

Substituting (6.31) and (6.33) into (6.30), then

$$\sum_{i=0}^{\gamma} \left[\overline{V} \right] = \sum_{i=0}^{\gamma} \left[\frac{1}{2} f_i^{\top}(\chi_{i,k}) P_i f_i(\chi_{i,k}) + \frac{1}{2} \chi_{i\delta,k+1}^{\top} P_i \chi_{i\delta,k+1} \right.$$

$$- \frac{1}{2} \chi_{i,k}^{\top} P_i \chi_{i,k} - \frac{1}{2} \chi_{i\delta,k}^{\top} P_i \chi_{i\delta,k}$$

$$- \frac{1}{4} P_{i,1}^{\top}(\chi_{i,k}, \chi_{i\delta,k})(R + P_{i,2}(\chi_{i,k}))^{-1} P_{i,1}(\chi_{i,k}, \chi_{i\delta,k})$$

$$+ \frac{1}{2} \| P_i \| \| f_i(\chi_{i,k}) \|^2 + \frac{1}{2} \| P_i \| \| \chi_{i\delta,k+1} \|^2$$

$$\left. + \frac{1}{2} \| P_i \| \| \chi_{i,k} \|^2 + \frac{1}{2} \| P_i \| \| \chi_{i\delta,k} \|^2 \right]. \tag{6.34}$$

In order to achieve asymptotic stability, it is required that $\overline{V} \leq 0$, then from (6.34) the next inequality is formulated

$$\sum_{i=0}^{\gamma} \left[\frac{1}{2} f_i^{\top}(\chi_{i,k}) P_i f_i(\chi_{i,k}) + \frac{1}{2} \chi_{i\delta,k+1}^{\top} P_i \chi_{i\delta,k+1} \right.$$

$$- \frac{1}{2} \chi_{i,k}^{\top} P_i \chi_{i,k} - \frac{1}{2} \chi_{i\delta,k}^{\top} P_i \chi_{i\delta,k}$$

$$\left. - \frac{1}{4} P_{i,1}^{\top}(\chi_{i,k}, \chi_{i\delta,k})(R + P_{i,2}(\chi_{i,k}))^{-1} P_{i,1}(\chi_{i,k}, \chi_{i\delta,k}) \right]$$

$$\leq \sum_{i=0}^{\gamma} \left[-\frac{1}{2} \| P_i \| \| f_i(\chi_{i,k}) \|^2 + \frac{1}{2} \| P_i \| \| \chi_{i\delta,k+1} \|^2 \right.$$

$$
-\frac{1}{2}\|P_i\|\|\chi_{i,k}\|^2 + \frac{1}{2}\|P_i\|\|\chi_{i\delta,k}\|^2 \Bigg]. \tag{6.35}
$$

Hence, selecting P_i such that (6.35) is satisfied, system (2.23) with control law (6.16) guarantees asymptotic trajectory tracking along the desired trajectory $\chi_{i\delta,k}$. It is worth noting that P_i and R_i are positive definite and symmetric matrices; thus, the existence of the inverse in (6.16) is ensured.

When function $-l_i(\chi_{i,k})$ is set to be the (6.30) right-hand side, that is,

$$
\begin{aligned}
\sum_{i=0}^{\gamma}\big[l_i(\chi_{i,k})\big] :=& -\sum_{i=0}^{\gamma}\big[\overline{V}_i\big]\big|_{u_{i,k}^*=\alpha_i(z_{i,k})} \\
=& \sum_{i=0}^{\gamma}\Bigg[\frac{1}{2}f_i^{\top}(\chi_{i,k})P_i f_i(\chi_{i,k}) + \frac{1}{2}\chi_{i\delta,k+1}^{\top}P_i\chi_{i\delta,k+1} \\
& -\frac{1}{2}\chi_{i,k}^{\top}P_i\chi_{i,k} - \frac{1}{2}\chi_{i\delta,k}^{\top}P_i\chi_{i\delta,k} - f_i^{\top}(\chi_{i,k})P_i\chi_{i\delta,k+1} \\
& -\frac{1}{4}P_{i,1}^{\top}(\chi_{i,k},\chi_{i\delta,k})(R+P_{i,2}(\chi_{i,k}))^{-1}P_{i,1}(\chi_{i,k},\chi_{i\delta,k}) \\
& +\chi_{i,k}^{\top}P_i\chi_{i\delta,k}\Bigg]
\end{aligned} \tag{6.36}
$$

then $V_i(\chi_{i,k})$ as proposed in (6.11) is a solution of the DT HJB equation (6.8). In order to obtain the optimal value for the cost functional (6.20), we substitute $l_i(\chi_{i,k})$ given in (6.36) into (6.20); then

$$
\begin{aligned}
\sum_{i=0}^{\gamma}\big[\mathscr{J}_i(z_{i,k})\big] =& \sum_{i=0}^{\gamma}\Bigg[\sum_{k=0}^{\infty}\big(l_i(z_{i,k})+u_{i,k}^{\top}R_i(z_{i,k})u_{i,k}\big)\Bigg] \\
=& \sum_{i=0}^{\gamma}\Bigg[\sum_{k=0}^{\infty}\big(-\overline{V}+u_{i,k}^{\top}R_i(z_{i,k})u_{i,k}\big)\Bigg] \\
=& \sum_{i=0}^{\gamma}\Bigg[-\sum_{k=0}^{\infty}\Bigg[\frac{1}{2}f_i^{\top}(\chi_{i,k})P_i f_i(\chi_{i,k}) + \frac{1}{2}\chi_{i\delta,k+1}^{\top}P_i\chi_{i\delta,k+1} \\
& -\frac{1}{2}\chi_{i,k}^{\top}P_i\chi_{i,k} - \frac{1}{2}\chi_{i\delta,k}^{\top}P_i\chi_{i\delta,k} + \chi_{i,k}^{\top}P_i\chi_{i\delta,k} \\
& -\frac{1}{4}P_{i,1}^{\top}(\chi_{i,k},\chi_{i\delta,k})(R+P_{i,2}(\chi_{i,k}))^{-1}P_{i,1}(\chi_{i,k},\chi_{i\delta,k}) \\
& -f_i^{\top}(\chi_{i,k})P_i\chi_{i\delta,k+1}\Bigg] + \sum_{k=0}^{\infty}u_{i,k}^{\top}R_i(z_{i,k})u_{i,k}\Bigg]
\end{aligned} \tag{6.37}
$$

Factorizing (6.37), and then adding the identity matrix

$$I_{i,m} = (R_i + P_{i,2}(\chi_{i,k}))(R_i + P_{i,2}(\chi_{i,k}))^{-1} \qquad (6.38)$$

with $I_{m_i} \in \Re^{m_i \times m_i}$, we obtain

$$
\begin{aligned}
\sum_{i=0}^{\gamma} \left[V_i(\chi_{i,k}) \right] = \sum_{i=0}^{\gamma} \Bigg[&-\sum_{k=0}^{\infty} \Bigg[\frac{1}{2} f_i^{\top}(\chi_{i,k}) P_i f_i(\chi_{i,k}) \\
&+ \frac{1}{2}\chi_{i\delta,k+1}^{\top} P_i \chi_{i\delta,k+1} - \frac{1}{2}\chi_{i,k}^{\top} P_i \chi_{i,k} - \frac{1}{2}\chi_{i\delta,k}^{\top} P_i \chi_{i\delta,k} \\
&- \frac{1}{4} P_{i,1}^{\top}(\chi_{i,k}, \chi_{i\delta,k})(R + P_{i,2}(\chi_{i,k}))^{-1}(R + P_{i,2}(\chi_{i,k})) \\
&\times (R + P_{i,2}(\chi_{i,k}))^{-1} P_{i,1}(\chi_{i,k}, \chi_{i\delta,k}) - f_i^{\top}(\chi_{i,k}) P_i \chi_{i\delta,k+1} \\
&+ \chi_{i,k}^{\top} P_i \chi_{i\delta,k} \Bigg] \Bigg] + \sum_{i=0}^{\gamma} \left[\sum_{k=0}^{\infty} u_{i,k}^{\top} R_i(z_{i,k}) u_{i,k} \right]
\end{aligned}
\qquad (6.39)
$$

Being $\alpha_i(\chi_{i,k}) = -\frac{1}{2}(R_i + P_{i,2}(\chi_{i,k}))^{-1} P_{i,1}(\chi_{i,k}, \chi_{i\delta,k})$, then (6.39) becomes

$$
\begin{aligned}
\sum_{i=0}^{\gamma} \left[V_i(\chi_{i,k}) \right] = \sum_{i=0}^{\gamma} \Bigg[&-\sum_{k=0}^{\infty} \Bigg[\frac{1}{2} f_i^{\top}(\chi_{i,k}) P_i f_i(\chi_{i,k}) \\
&+ \frac{1}{2}\chi_{i\delta,k+1}^{\top} P_i \chi_{i\delta,k+1} - f_i^{\top}(\chi_{i,k}) P_i \chi_{i\delta,k+1} + \chi_{i,k}^{\top} P_i \chi_{i\delta,k} \\
&- \frac{1}{2}\chi_{i,k}^{\top} P_i \chi_{i,k} - \frac{1}{2}\chi_{i\delta,k}^{\top} P_i \chi_{i\delta,k} + P_{i,1}^{\top}(\chi_{i,k}, \chi_{i\delta,k})\alpha(z_{i,k}) \\
&+ \alpha^{\top}(z_{i,k}) P_{i,2} \alpha(z_{i,k}) + \alpha^{\top}(z_{i,k}) R_i \alpha(z_{i,k}) \Bigg] \\
&+ \sum_{k=0}^{\infty} u_{i,k}^{\top} R_i(z_{i,k}) u_{i,k} \Bigg] \\
= \sum_{i=0}^{\gamma} \Bigg[&-\sum_{k=0}^{\infty} \Bigg[\frac{1}{2}(\chi_{i,k+1} - \chi_{i\delta,k+1})^{\top} P_i (\chi_{i,k+1} - \chi_{i\delta,k+1}) \\
&- \frac{1}{2}(\chi_{i,k} - \chi_{i\delta,k})^{\top} P_i (\chi_{i,k} - \chi_{i\delta,k}) + u_{i,k}^{*\top} R_i u_{i,k}^{*} \Bigg] \\
&+ \sum_{k=0}^{\infty} u_{i,k}^{\top} R_i(z_{i,k}) u_{i,k} \Bigg] \\
= \sum_{i=0}^{\gamma} \Bigg[&-\sum_{k=0}^{\infty} \left[V_i(z_{i,k+1}) - V_i(z_{i,k}) \right] \Bigg] \\
&+ \sum_{i=0}^{\gamma} \left[\sum_{k=0}^{\infty} \left[u_{i,k}^{\top} R_i(z_{i,k}) u_{i,k} - \alpha^{\top}(z_{i,k}) R_i \alpha(z_{i,k}) \right] \right]
\end{aligned}
\qquad (6.40)
$$

which can be written as

$$
\sum_{i=0}^{\gamma} \big[V_i(z_{i,k}) \big] = \sum_{i=0}^{\gamma} \Bigg[- \sum_{k=1}^{\infty} \big[V_i(z_{i,k+1}) - V_i(z_{i,k}) \big] - V_i(z_{i,1})
$$

$$
+ V_i(z_{i,0}) + \sum_{k=0}^{\infty} \Big[u_{i,k}^\top R_i(z_{i,k}) u_{i,k}
$$

$$
- \alpha_i^\top(z_{i,k}) R_i(z_{i,k}) \alpha_i(z_{i,k}) \Big] \Bigg]
$$

$$
= \sum_{i=0}^{\gamma} \Bigg[- \sum_{k=2}^{\infty} \big[V_i(z_{i,k+1}) - V_i(z_{i,k}) \big] - V_i(z_{i,2})
$$

$$
+ V_i(z_{i,1}) - V_i(z_{i,1}) + V_i(z_{i,0}) \Bigg] \tag{6.41}
$$

$$
+ \sum_{i=0}^{\gamma} \Bigg[+ \sum_{k=0}^{\infty} \big[u_{i,k}^\top R_i(z_{i,k}) u_{i,k} - \alpha_i^\top(z_{i,k}) R_i(z_{i,k}) \alpha_i(z_{i,k}) \big] \Bigg]
$$

For notation convenience in (6.41), the upper limit ∞ will treated as $N \to \infty$, and thus

$$
\sum_{i=0}^{\gamma} \big[\mathscr{V}_i(z_{i,k}) \big] = \sum_{i=0}^{\gamma} \big[-V_i(z_{i,N})
$$

$$
+ V_i(z_{i,N-1}) - V_i(z_{i,N-1}) + V_i(z_{i,0})
$$

$$
+ \lim_{N \to \infty} \sum_{k=0}^{N} \big[u_{i,k}^\top R_i(z_{i,k}) u_{i,k} - \alpha_i^\top(z_{i,k}) R_i(z_{i,k}) \alpha_i(z_{i,k}) \big] \Bigg]
$$

$$
= \sum_{i=0}^{\gamma} \Bigg[-V_i(z_{i,N}) + V_i(z_{i,0}) \tag{6.42}
$$

$$
+ \lim_{N \to \infty} \sum_{k=0}^{N} \big[u_{i,k}^\top R_i(z_{i,k}) u_{i,k} - \alpha_i^\top(z_{i,k}) R_i(z_{i,k}) \alpha_i(z_{i,k}) \big] \Bigg]
$$

Letting $N \to \infty$ and noting that $V_i(z_{i,N}) \to 0$ for all $z_{i,0}$, then

$$
\sum_{i=0}^{\gamma} \big[V_i(z_{i,k}) \big] = \sum_{i=0}^{\gamma} \Bigg[V_i(z_{i,0}) + \sum_{k=0}^{\infty} \big[u_{i,k}^\top R_i(z_{i,k}) u_{i,k} - \alpha_i^\top(z_{i,k}) R_i(z_{i,k}) \alpha_i(z_{i,k}) \big] \Bigg] \tag{6.43}
$$

Thus, the minimums values of (6.43) are reached with $\sum_{i=0}^{\gamma} [u_{i,k}] = \sum_{i=0}^{\gamma} [\alpha_i(z_{i,k})]$. Hence, the control laws (6.16) minimizes the cost functional (6.20). The optimal value function of (6.20) is $\sum_{i=0}^{\gamma} [\mathscr{J}_i^*(z_{i,0})] = \sum_{i=0}^{\gamma} [V_i(z_{i,0})]$ for all $z_{i,0}$. ■

Optimal control will be in general of the form (6.11) and the minimum value of the performance index will be function of the initial state $z_{i,0}$. If system (6.10) and the control law (6.11) are given, we shall say that the pair $\{V_i(z_{i,k}), l_i(z_{i,k})\}$ is a solution to the *inverse optimal control problem* if the performance index (6.1) is minimized by (6.11), and the minimum value being $V_i(z_{i,0})$.

6.3 Super-Twisting Observer

Let consider the nonlinear system in the sate-space form

$$\theta_{k+1,1} = \theta_{k,1} + T\theta_{k,2}$$
$$\theta_{k+1,2} = \theta_{k,2} + Tf(k, \theta_{k,1}, \theta_{k,2}, u_k) + \xi(k, \theta_{k,1}, \theta_{k,2}, u_k)$$
$$u = U(k, \theta_{k,1}, \theta_{k,2})$$
$$y_k = \theta_{k,1} \tag{6.44}$$

where the nominal part of the system dynamics is represented by the function $f(k, \theta_{k,1}, \theta_{k,2}, u_k)$, while the uncertainties are concentrated in the term $\xi(k, \theta_{k,1}, \theta_{k,2}, u_k)$. The state $\theta_{k,2}$ of system (6.44) is to be observed, while only the state $\theta_{k,1}$ is available. Only the scalar case $\theta_{k,1}; \theta_{k,2} \in \Re$ is considered for the sake of simplicity [4]. In the vector case, the observers are constructed in parallel for each position variable $\theta_{1,j}$ on exactly the same way. The super-twisting observer has the form

$$\hat{\theta}_{k+1,1} = \hat{\theta}_{k,1} + T\hat{\theta}_{k,2} + v_{k,1}$$
$$\hat{\theta}_{k+1,2} = \hat{\theta}_{k,2} + Tf(k, \theta_{k,1}, \hat{\theta}_{k,2}, u_k)$$
$$+ \xi(k, \theta_{k,1}, \theta_{k,2}, u_k) + v_{k,2}$$
$$y_k = \theta_{k,1} \tag{6.45}$$

where $\dot{\hat{\theta}}_{k,1}$ and $\dot{\hat{\theta}}_{k,2}$ are the state estimations, and the correction variables $v_{k,1}$ and $v_{k,2}$ are output injections of the form

$$v_{k,1} = -k_1 \|\theta_{k,1} - \hat{\theta}_{k,1}\|^{1/2} sign(\theta_{k,1} - \hat{\theta}_{k,1})$$
$$v_{k,2} = -k_2 sign(\theta_{k,1} - \hat{\theta}_{k,1}). \tag{6.46}$$

It is taken from the definition that at the initial moment $\hat{\theta}_{k,1} = \theta_{k,1}$ and $\hat{\theta}_{k,2} = 0$. Defining $\tilde{\theta}_{k,1} = \theta_{k,1} - \hat{\theta}_{k,1}$ and $\tilde{\theta}_{k,2} = \theta_{k,2} - \hat{\theta}_{k,2}$ we obtain the error equations

$$\dot{\tilde{\theta}}_{k,1} = \tilde{\theta}_{k,2} - k_1 \|\tilde{\theta}_{k,1}\|^{1/2} sign(\tilde{\theta}_{k,1})$$
$$\tilde{\theta}_{k+1,1} = F(k, \theta_{k,1}, \theta_{k,2}, \hat{\theta}_{k,2}) - k_2 sign(\tilde{\theta}_{k,1}). \tag{6.47}$$

where $F(k, \theta_{k,1}, \theta_{k,2}, \hat{\theta}_{k,2}) = f(k, \theta_{k,1}, \theta_{k,2}, U(k, \theta_{k,1}, \theta_{k,2})) - f(k, \theta_{k,1}, \hat{\theta}_{k,2}, U(k, \theta_{k,1}, \theta_{k,2})) + \xi(k, \theta_{k,1}, \theta_{k,2}, U(t, \theta_{k,1}, \theta_{k,2}))$. Suppose that the system states can be assumed bounded, then the existence is ensured of a constant f^+ [3].

References

1. Al-Tamimi, A., Lewis, F.L., Abu-Khalaf, M.: Discrete-time nonlinear HJB solution using approximate dynamic programming: convergence proof. IEEE Trans. Syst. Man Cybern. Part B Cybern. **38**(4), 943–949 (2008)
2. Basar, T., Olsder, G.J.: Dynamic Noncooperative Game Theory. Academic Press, New York (1995)
3. Davila, J., Fridman, L., Levant, A.: Second-Order sliding mode observer for mechanical systems. IEEE Trans. Autom. Control **50**(11), 1785–1789 (2005)
4. Davila, J., Fridman, L., Poznyak, A.: Observation and identification of mechanical systems via second order sliding modes. Int. J. Control **79**(10), 1251–1262 (2006)
5. Haddad, W.M., Chellaboina, V.-S., Fausz, J.L., Abdallah, C.: Identification and control of dynamical systems using neural networks. J. Frankl. Inst. **335**(5), 827–839 (1998)
6. Kirk, D.E.: Optimal Control Theory: An Introduction. Dover Publications Inc, Englewood Cliffs (2004)
7. Lewis, F.L., Syrmos, V.L.: Optimal Control. Wiley, New York (1995)
8. Ohsawa, T., Bloch, A.M., Leok, M.: Discrete Hamilton-Jacobi theory and discrete optimal control. In: Proceedings of the 49th IEEE Conference on Decision and Control, pp. 5438–5443. Atlanta (2010)
9. Sanchez, E.N., Ornelas-Tellez, F.: Discrete-time inverse optimal control for nonlinear systems. CRC Press, Boca Raton (2013)
10. Sepulchre, R., Jankovic, M., Kokotovic, P.V.: Constructive Nonlinear Control. Springer, London (1997)

Chapter 7
Robotics Application

7.1 Experimental Prototypes

In order to evaluate the performance of the proposed control algorithms, five prototypes were used:

- Two DOF robot manipulator
- Five DOF redundant robot
- Seven DOF Mitsubishi PA10-7CE robot arm
- KUKA youBot mobile robot
- Shrimp mobile robot.

7.1.1 Two DOF Robot Manipulator

This robot manipulator consists of two rigid links; high-torque brushless direct-
drive servos are used to drive the joints without gear reduction. This kind of
joints present reduced backlash and significantly lower joint friction as compared
to the actuators with gear drives. The motors used in the experimental arm are
DM1200-A and DM1015-B from Parker Compumotor, for the shoulder and elbow
joints, respectively. Angular information is obtained from incremental encoders
located on the motors, which have a resolution of 1,024,000 pulses/rev for the first
motor and 655,300 for the second one (accuracy 0.0069° for both motors), and the
angular velocity information is computed via numerical differentiation of the angular
position signal. The two DOF robot manipulator is shown in Figs. 7.1 and 7.2 and
its respective numeric values are included in Table 7.1 [10].

© Springer International Publishing Switzerland 2017   69
R. Garcia-Hernandez et al., *Decentralized Neural Control: Application to Robotics*,
Studies in Systems, Decision and Control 96, DOI 10.1007/978-3-319-53312-4_7

Fig. 7.1 Two DOF robot manipulator

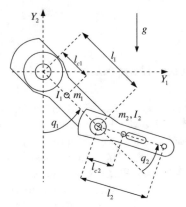

Fig. 7.2 Axes of two DOF robot manipulator

Table 7.1 Robot manipulator parameters

Link	Description	Notation	Value	Unit
(1) Shoulder	Mass link 1	m_1	23.902	kg
	Length link 1	l_1	0.450	m
	Inertia link 1	I_1	1.266	kg m^2
	Center of mass link 1	l_{c1}	0.091	m
(2) Elbow	Mass link 2	m_2	3.880	kg
	Length link 2	l_2	0.450	m
	Inertia link 2	I_2	0.093	kg m^2
	Center of mass link 2	l_{c2}	0.048	m

7.1.2 Five DOF Redundant Robot

A view of the modular Articulated Nimble Adaptable Trunk (ANAT) robot is shown in Fig. 7.3, and its different axes are illustrated in Fig. 7.4. In this patented design, each joint module contains a direct current (DC) motor component, sealed axes bearings, drive output position transducers, and a controller, all integrated into a lightweight aluminum structure. Each self-contained joint module is joined to its adjacent modules by screws. Two adjacent modules can be mechanically decoupled facilitating maintenance or retrofit. These modules are grouped together to form a stair-like shape, which handles a substantial load. This modular construction allows for manipulator configuration covering a broad range of sizes, payloads, and kinematics configurations for different applications.

The module incorporates an advanced, fully digital distributed servo control system, containing a micro controller. All necessary servo electronics (communication, data acquisition, PWM amplifier, computing, etc.) are co-located in each module. This distributed hardware architecture eliminates the complex arm harness previously required, reduces it to a single power supply and one cable to communicate with an external PC or internal Digital Signal Processor (DSP). This on-board electronics system is particularly valuable for mobile vehicle installations, since the size

Fig. 7.3 ANAT redundant robot

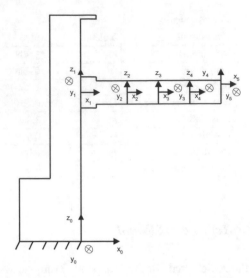

Fig. 7.4 ANAT axes

of the control cabinet is dramatically reduced. Furthermore, this on-board electronics system allows the module to be fully independently controlled.

The previously described design becomes evident for the need of innovative compound serial manipulators and robotic arms for manufacturing, or manipulating tools; serpentine in its flexibility, it exceeds human dexterity and allows accessing confined spaces and tortuous paths. Scalability from small to large diameters makes it adaptable to many tasks. Modularity increases the fault tolerance of ANAT robot by providing redundant yaw axes. If one of the modules fails, the next modules would be able to provide sufficient mobility to complete the mission. This innovative technology opens the door to many applications such as: telerobotics, spray finishing, manipulating, cutting, welding, grinding, as well as prospecting. This same technology allows using the robot as a fixed manipulator or a mobile robot. The numeric values for workspace of ANAT redundant robot are included in Table 7.2.

Table 7.2 Workspace ANAT redundant robot

Link	Type	Workspace	Unit
1	Prismatic	$L_0 = 0.57$ to 1.27	m
2	Rotative	$-\pi/2$ to $\pi/2$	rad
3	Rotative	$-\pi/2$ to $\pi/2$	rad
4	Rotative	$-\pi/2$ to $\pi/2$	rad
5	Rotative	$-\pi/2$ to $\pi/2$	rad

7.1.3 Seven DOF Mitsubishi PA10-7CE Robot Arm

The Mitsubishi PA10-7CE arm is an industrial robot manipulator which completely changes the vision of conventional industrial robots. Its name is an acronym of Portable General-Purpose Intelligent Arm. There exist two versions [4]: the PA10-6C and the PA10-7C, where the suffix digit indicates the number of degrees of freedom of the arm. This work focuses on the study of the PA10-7CE model, which is the enhanced version of the PA10-7C. The PA10 arm is an open architecture robot; it means that it possesses [15]:

- A hierarchical structure with several control levels.
- Communication between levels, via standard interfaces.
- An open general purpose interface in the higher level.

This scheme allows the user to focus on the programming of the tasks at the PA10 system higher level, without regarding on the operation of the lower levels. The programming can be performed using a high level language, such as Visual BASIC or Visual C++, from a PC with Windows operating system. The PA10 robot is currently the open architecture robot more employed for research (see, e.g., [7, 11, 16]). The PA10 system is composed of four sections or levels, which conform a hierarchical structure [13]:

Level 4: Operation control section (OCS); formed by the PC and the teaching pendant.
Level 3: Motion control section (MCS); formed by the motion control and optical boards.

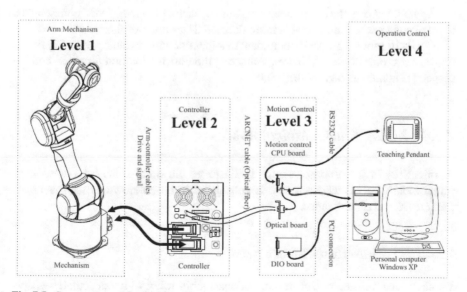

Fig. 7.5 Components of the PA10 system

Fig. 7.6 Mitsubishi
PA10-7CE robot axes

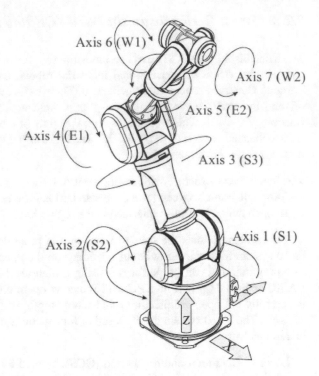

Level 2: Servo drives.
Level 1: Robot manipulator.

Figure 7.5 shows the relation existing among each of the mentioned components. The following subsections give a more detailed description of them.

The PA10 robot is a 7-DOF redundant manipulator with revolute joints. Figure 7.6 shows a diagram of the PA10 arm, indicating the positive rotation direction and the respective names of each of the joints.

7.1.4   KUKA YouBot Mobile Robot

Applicability of the propose scheme is illustrated via simulations for a omnidirectional mobile robot, whose model is considered to be unknown as well as all its parameters and disturbances.

7.1.4.1   Omnidirectional Mobile Robot

We consider a KUKA youBot omnidirectional mobile robot with four actuated wheels as shown in Fig. 7.7. For each actuated wheel, a suitable controller is designed accord-

Fig. 7.7 KUKA youBot

ing to the described decentralized scheme. The actuated wheels are directly driven by DC motors. The dynamics of each DC motor can be expressed by the following state-space model [1]:

$$\chi_{i,k+1}^1 = \chi_{i,k}^1 + T\left(-\frac{b}{J}\chi_{i,k}^1 + \frac{K_t}{J}\chi_{i,k}^2\right) + \sum_{\ell=1,\ \ell\neq i}^{\gamma} \Gamma_{1\ell,k}(\chi_\ell)$$

$$\chi_{i,k+1}^2 = \chi_{i,k}^2 + T\left(-\frac{K_b}{L_a}\chi_{i,k}^1 - \frac{R_a}{L_a}\chi_{i,k}^2 + \frac{1}{L_a}u_{i,k}\right) + \sum_{\ell=1,\ \ell\neq i}^{\gamma} \Gamma_{2\ell,k}(\chi_\ell) \quad (7.1)$$

where χ_i^1 represent the angular velocity in $\frac{rad}{s}$ for each motor respectively with $i = 1, \ldots, 4$. Accordingly, each actuated wheel is considered as a subsystem. χ_i^2 is the armature current in Amp. $\Gamma_{i\ell,k}(\chi_\ell)$ reflect the interaction between the i-th and the ℓ-th subsystem with $1 \leq \ell \leq \gamma$. The input terminal voltage u_i is taken to be the controlling variable. The sampling time is selected as $T = 0.01$ s. The parameters of DC motor are included in Table 7.3.

7.1.4.2 Neural Identification

Each wheel DC motor is driving whose model is approximated by a neural identifier as developed in Sect. 5.2. To obtain the discrete-time neural model for the electrically drives of KUKA youBot robot (7.1), the following neural identifier trained with EKF (5.26) is proposed

Notation	Description	Value	Unit
R_a	Electric resistance	0.6	Ohms
L_a	Armature inductance	0.012	H
K_t	Torque factor constant	0.8	Nm/Amp
K_b	Back emf constant	0.8	Vs/rad
J	Moment of inertia	0.0167	Kgm^2/s^2
b	Viscous friction constant	0.0167	Nms

Table 7.3 Parameters of DC motor for KUKA mobile robot

$$x_{i,k+1}^1 = w_{11i,k}S(\chi_{i,k}^1) + w_{12i,k}S(\chi_{i,k}^1) + w_{1i}'\chi_{2i,k}$$
$$x_{i,k+1}^2 = w_{21i,k}S(\chi_{i,k}^2) + w_{22i,k}S(\chi_{i,k}^1) + w_{23i,k}S(\chi_{i,k}^2) + w_{2i}'u_{i,k} \qquad (7.2)$$

where x_i^1 and x_i^2 identify the motor angular velocities χ_i^1 and the respective motor current χ_i^2. The NN training is performed on-line, and all of its states are initialized, randomly. The RHONN parameters are heuristically selected as:

$$P_{iq}^1(0) = 1 \times 10^{10}I \quad R_{iq}^1(0) = 1 \times 10^8 \quad w_{1i}' = 0.001$$
$$P_{iq}^2(0) = 1 \times 10^{10}I \quad R_{iq}^2(0) = 5 \times 10^3 \quad w_{2i}' = 1$$
$$Q_{iq}^1(0) = 1 \times 10^7I \quad Q_{iq}^2(0) = 5 \times 10^3I$$

where I is the identity matrix of adequate dimension. It is important to consider that for the EKF-learning algorithm the covariance matrices are used as design parameters [2, 3]. The neural network structure (7.2) is determined heuristically in order to minimize the state estimation error. It is important to remark that the initial conditions of the plant are completely different from the initial conditions of the neural network.

7.1.4.3   Control Law Implementation

The control law proposed for this application takes the following form:

$$f_i(x_{i,k}) = \begin{bmatrix} w_{11i,k}S(\chi_{i,k}^1) + w_{12i,k}S(\chi_{i,k}^1) + w_{1i}'\chi_{2i,k} \\ w_{21i,k}S(\chi_{i,k}^2) + w_{22i,k}S(\chi_{i,k}^1) + w_{23i,k}S(\chi_{i,k}^2) \end{bmatrix} \qquad (7.3)$$

and

$$g_i(x_{i,k}) = \begin{bmatrix} 0 \\ w_{2i}' \end{bmatrix} \qquad (7.4)$$

where $P_{i,1}(x_{i,k}) = g_i^\top(x_{i,k}) P_i f_i(x_{i,k})$ and $P_{i,2}(x_{i,k}) = \frac{1}{2} g_i^\top(x_{i,k}) P_i g_i(x_{i,k})$. The value of P is equal for each subsystem, thus

$$P_i = \begin{bmatrix} 1.1 & 0.11 \\ 0.11 & 1.1 \end{bmatrix} \tag{7.5}$$

where $u_{i,k}$ is as follows:

$$u_{i,k} = -\frac{1}{2}(R_i(x_{i,k}) + P_{i,2}(x_{i,k}))^{-1} P_{i,1}(x_{i,k}) \tag{7.6}$$

7.1.5 Shrimp Mobile Robot

Applicability of the scheme is illustrated via real-time implementation for a rough terrain mobile robot, whose model is considered to be unknown as well as all its parameters and disturbances.

7.1.5.1 Mobile Robot

We consider a Shrimp mobile robot with eighth actuated wheels as shown in Fig. 7.8. The dynamic of a DC motor can be expressed in the following state-space model [1]:

Fig. 7.8 Shrimp – on rough terrain

Table 7.4 Parameters of DC motor for Shrimp mobile robot

Notation	Description	Value	Unit
R_a	Electric resistance	0.6	Ohms
L_a	Armature inductance	0.012	H
K_t	Torque factor constant	0.8	Nm/Amp
K_b	Back emf constant	0.8	Vs/rad
J	Moment of inertia	0.0167	Kgm^2/s^2
b	Viscous friction constant	0.0167	Nms

$$\chi_{i,k+1}^1 = \chi_{i,k}^1 + T\left(-\frac{b}{J}\chi_{i,k}^1 + \frac{K_t}{J}\chi_{i,k}^2\right)$$

$$\chi_{i,k+1}^2 = \chi_{i,k}^2 + T\left(-\frac{K_b}{L_a}\chi_{i,k}^1 - \frac{R_a}{L_a}\chi_{i,k}^2 + \frac{1}{L_a}u_{i,k}\right) \tag{7.7}$$

where χ_i^1 represent the angular velocity in $\frac{\text{rad}}{\text{s}}$ for each motor respectively with $i = 1, \ldots, 8$. Accordingly, each actuated wheel is considered as a subsystem. χ_i^2 is the armature current in Amp. The input terminal voltage u_i is taken to be the controlling variable. The sampling time is selected as $T = 0.01\,\text{s}$. The parameters of DC motor are included in Table 7.4.

7.1.5.2   Super-Twisting Observer Synthesis

The proposed super-twisting observer has the form

$$\hat{x}_{i,k+1}^1 = \chi_{i,k}^1 + T\left(-\frac{b}{J}\chi_{i,k}^1 + \frac{K_t}{J}\hat{x}_{i,k}^2\right) + v_{i,k}^1 \tag{7.8}$$

$$\hat{x}_{i,k+1}^2 = \hat{x}_{i,k}^2 + T\left(-\frac{K_b}{L_a}\chi_{i,k}^1 - \frac{R_a}{L_a}\hat{x}_{i,k}^2 + \frac{1}{L_a}u_{i,k}\right) + v_{i,k}^2$$

where $\hat{x}_{i,k+1}^1$ and $\hat{x}_{i,k+1}^2$ are the state estimations of the angular velocities χ_i^1 and the motor currents χ_i^2, respectively. The correction variables $v_{i,k}^1$ and $v_{i,k}^2$ are output injections of the form

$$v_{i,k}^1 = -k_1\|\chi_{i,k}^1 - \hat{x}_{i,k}^1\|^{1/2}sign(\chi_{i,k}^1 - \hat{x}_{i,k}^1)$$

$$v_{i,k}^2 = -k_2 sign(\chi_{i,k}^1 - \hat{x}_{i,k}^1). \tag{7.9}$$

where the constant gains are taken as $k_1 = 0.01$ and $k_2 = 0.001$.

7.1.5.3 Neural Identification Synthesis

We apply the neural identifier, developed in Sect. 5.2, to obtain a discrete-time neural model for the electrically drives of the Shrimp robot (7.8) which is trained with the EKF (5.26) respectively, as follows:

$$x_{i,k+1}^1 = w_{11i,k} S(\chi_{i,k}^1) + w_{12i,k} S(\chi_{i,k}^1) + w_{1i}' \hat{x}_{i,k}^2$$
$$x_{i,k+1}^2 = w_{21i,k} S(\hat{x}_{i,k}^2) + w_{22i,k} S(\chi_{i,k}^1) + w_{23i,k} S(\hat{x}_{i,k}^2) + w_{2i}' u_{i,k} \qquad (7.10)$$

where x_i^1 and x_i^2 identify the angular velocities χ_i^1 and the motor currents χ_i^2, respectively. The NN training is performed on-line, and all of its states are initialized, randomly. The RHONN parameters are heuristically selected as:

$$
\begin{aligned}
P_{iq}^1(0) &= 1 \times 10^{10} I & R_{iq}^1(0) &= 1 \times 10^8 & w_{1i}' &= 1 \\
P_{iq}^2(0) &= 1 \times 10^{10} I & R_{iq}^2(0) &= 5 \times 10^3 & w_{2i}' &= 1 \\
Q_{iq}^1(0) &= 1 \times 10^7 I & Q_{iq}^2(0) &= 5 \times 10^3 I &
\end{aligned}
$$

where I is the identity matrix. It is important to consider that for the EKF-learning algorithm the covariances are used as design parameters [2, 3]. The neural network structure (7.10) is determined heuristically in order to minimize the state estimation error.

7.1.5.4 Control Synthesis

The goal is to force the state $x_{i,k}^1$ to track a desired reference signal $\chi_{i\delta,k}^1$, which is achieved by a control law as described in Sect. 6.2. First the tracking error is defined as

$$z_{i,k}^1 = x_{i,k}^1 - \chi_{i\delta,k}^1$$

Then using (7.10) and introducing the desired dynamics for $z_{i,k}^1$ results in

$$z_{i,k+1}^1 = w_{i,k}^1 \varphi_{1i}(\chi_{i,k}^1) + w_{i,k}'^1 \chi_{i,k}^2 - \chi_{i\delta,k+1}^1$$
$$= K_i^1 z_{i,k}^1 \qquad (7.11)$$

where $|K_i^1| < 1$. The desired value $\chi_{i\delta,k}^2$ for the pseudo-control input $\chi_{i,k}^2$ is calculated from (7.11) as

$$\chi_{i\delta,k}^2 = \frac{1}{w_{i,k}'^1}(-w_{i,k}^1 \varphi_{1i}(\chi_{i,k}^1) + \chi_{i\delta,k+1}^1 + K_i^1 z_{i,k}^1) \qquad (7.12)$$

At the second step, we introduce a new variable as

$$z_{i,k}^2 = x_{i,k}^2 - \chi_{i\delta,k}^2$$

Taking one step ahead, we have

$$z_{i,k+1}^2 = w_{i,k}^2 \varphi_{2i}(\chi_{i,k}^1, \chi_{i,k}^2) + w_{i,k}^{\prime 2} u_{i,k} - \chi_{i\delta,k+1}^2 \tag{7.13}$$

where $u_{i,k}$ is defined as

$$u_{i,k} = -\frac{1}{2}\left(R_i(z_k) + g_i^\top(x_{i,k})P_i g_i(z_k)\right)^{-1} \times g_i^\top(x_{i,k})P_i(f_i(x_{i,k}) - x_{i\delta,k+1}) \tag{7.14}$$

The controllers parameters are shown below:

$$P_i = \begin{bmatrix} 1.6577 & 0.6299 \\ 0.6299 & 2.8701 \end{bmatrix}.$$

7.2 Real-Time Results

The real-time experiments are performed using the two DOF robot manipulator, the five DOF redundant robot and Shrimp mobile robot described in the previous sections. The implemented control structures include the decentralized neural identification and control scheme presented in Chap. 3, the decentralized neural backstepping approach analyzed in Chap. 4 and the decentralized inverse optimal control for trajectory tracking proposed in Chap. 6.

7.2.1 Case Study: Two DOF Robot Manipulator

For experiments, we select the following discrete-time trajectories

$$\begin{aligned} x_{1d,k}^1 &= b_1(1 - e^{d_1 kT^3}) + c_1(1 - e^{d_1 kT^3})\sin(\omega_1 kT)[\text{rad}] \\ x_{2d,k}^1 &= b_2(1 - e^{d_2 kT^3}) + c_2(1 - e^{d_2 kT^3})\sin(\omega_2 kT)[\text{rad}] \end{aligned} \tag{7.15}$$

where $b_1 = \pi/4$, $c_1 = \pi/18$, $d_1 = -2.0$, and $\omega_1 = 5$ [rad/s] are parameters of the desired position trajectory for the first joint, whereas $b_2 = \pi/3$, $c_2 = 25\pi/36$, $d_2 = -1.8$, and $\omega_2 = 1.0$ [rad/s] are parameters of the desired position trajectory for the second joint. The sampling time is selected as $T = 2.5$ ms.

These selected trajectories (7.15) present the following characteristics: (a) incorporate a sinusoidal term to evaluate the performance in presence of relatively fast periodic signals, from which the non-linearities of the robot dynamics are really important and (b) present a term which smoothly grows for maintaining the robot in an operation state without saturating actuators. The actuator limits are 150 and 15 [Nm], respectively.

7.2.1.1 Real-Time Decentralized Neural Block Control Results

Figure 7.9 displays the identification and trajectory tracking results for each joint. In the real system the initial conditions are the same that those of the neural identifier, both are restricted to be equal to zero in according to the experimental prototype architecture; therefore transient errors do not appear. The tracking errors for each joint are presented in Fig. 7.10.

The applied torques to each joint are shown in Fig. 7.11. The control signals present oscillations at some time instants due to gains and fixed parameters selected for each controller. It is easy to see that both control signals are always inside of the prescribed limits given by the actuators manufacturer, that is, their absolute values are smaller than the bounds τ_1^{max} and τ_2^{max}, respectively.

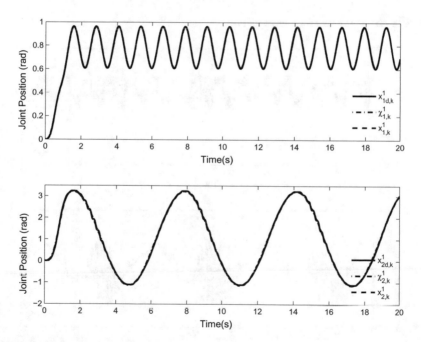

Fig. 7.9 Identification and tracking for joints 1 and 2

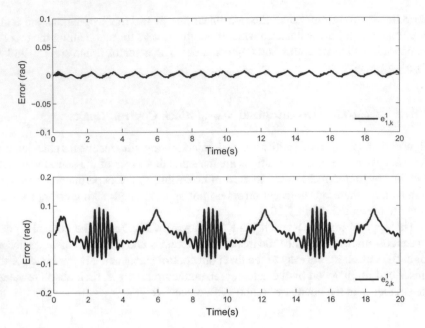

Fig. 7.10 Tracking errors for joints 1 and 2

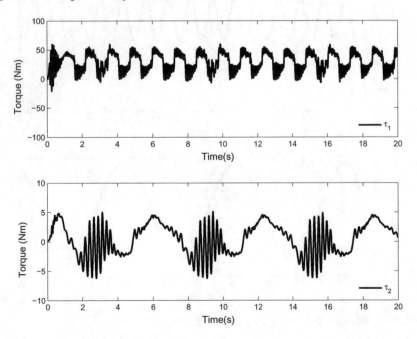

Fig. 7.11 Applied torques to joints 1 and 2

7.2.1.2 Real-Time Decentralized Neural Backstepping Control Results

The trajectory tracking results are presented in Fig. 7.12. The tracking performance can be verified for each of plant outputs, respectively. Figure 7.13 displays the angular position tracking errors for joints 1 and 2.

The applied torques to each joint are shown in Fig. 7.14. It is easy to see that both control signal are always inside of the prescribed limits given by the actuators manufacturer; that is, their absolute values are smaller than the limits τ_1^{max} and τ_2^{max}, respectively.

Time evolution of the position error $e_{i,k}^1$ reflects that the control system performance is very good. The performance criterion considered is the mean square error (MSE) value of the position error calculated as

$$\text{MSE}[e_{i,k}^1] = \sqrt{\frac{1}{t}\sum_{k=0}^{n}\|e_{i,k}^1\|^2 T} \tag{7.16}$$

where T is the sampling time and $t = 20$ s.

A comparative analysis of the two proposed schemes is included in Table 7.5. DNBC means Decentralized Neural Block Control and DNBS means Decentralized Neural Backstepping Control.

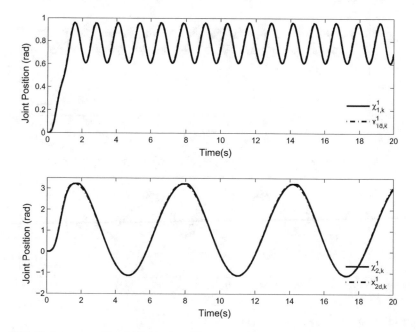

Fig. 7.12 Trajectory tracking for joints 1 and 2

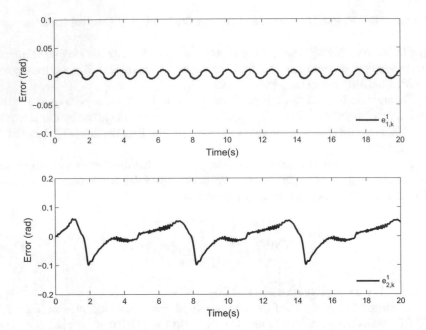

Fig. 7.13 Tracking errors for joints 1 and 2

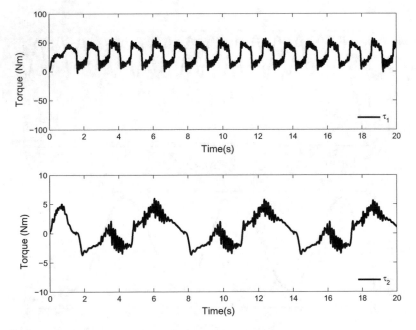

Fig. 7.14 Applied torques to joints 1 and 2

Table 7.5 Comparison of the MSE for the real joint positions 2-DOF robot

Algorithm	$e^1_{1,k}$	$e^1_{2,k}$
DNBC	8.3325c-6	0.0015
DNBS	4.1611e-5	0.0013

According to the mean square error presented in Table 7.5, the scheme with better performance is the one based on the block control and sliding mode technique; on the other hand the scheme with slightly lower performance is the one based on the backstepping technique. It is important to remark that both schemes present an adequate performance for the case of trajectory tracking. Both algorithms have a complexity of order $O(n^2)$, however computational requirements are lower for the backstepping technique due to the fact that the neural identifier within the control scheme is no longer need.

In the literature, there are decentralized control algorithms in continuous-time [9, 18], however the order of complexity is higher due to the used numerical method. Furthermore, regarding the complexity issue, discrete-time results are available only for SISO systems [6, 19]. Other control schemes can not be compared because there are interconnections involved which are not rather quantifiable [5, 8, 12, 14].

7.2.2 Case Study: Five DOF Robot Manipulator

We select the following discrete-time trajectories

$$
\begin{aligned}
x^1_{1\mathrm{d},k} &= a_1 \tanh(b_1 kT)[\mathrm{m}] \\
x^1_{2\mathrm{d},k} &= a_2 \sin(b_2 kT)[\mathrm{rad}] \\
x^1_{3\mathrm{d},k} &= -a_3 \sin(b_3 kT)[\mathrm{rad}] \\
x^1_{4\mathrm{d},k} &= a_4 \sin(b_4 kT)[\mathrm{rad}] \\
x^1_{5\mathrm{d},k} &= -a_5 \sin(b_5 kT)[\mathrm{rad}]
\end{aligned}
\tag{7.17}
$$

where $a_1 = 0.05$ and $b_1 = 0.1$ are parameters of the desired position trajectory for the prismatic joint, whereas $a_2 = a_3 = a_4 = a_5 = 45\pi/180$, $b_2 = b_3 = b_4 = b_5 = 0.5$ [rad/s] are parameters of the desired position trajectory for each rotative joint, respectively. The sampling time is selected as $T = 5$ ms.

These selected trajectories (7.17) incorporate a sinusoidal term to evaluate the performance in presence of relatively fast periodic signals, for which the non-linearities of the robot dynamics are really important.

7.2.2.1 Real-Time Decentralized Neural Block Control Results

Real-time results for identification and trajectory tracking using the decentralized
neural block control (DNBC) scheme are shown in Fig. 7.15. For the real system the
initial conditions are the same that those of the neural identifier, both restricted to be
equal to zero; therefore does not exist transient errors.

The tracking error performance for each joint are presented in Fig. 7.16. The
applied torques for each joint are shown in Fig. 7.17. The control signals present
oscillations at some time instants due to gains and fixed parameters selected for each
controller. It is easy to see that all control signals are always inside of the prescribed
limits given by the actuators manufacturer; that is, their absolute values are smaller
than the bounds τ_1^{max} to τ_5^{max}, respectively.

7.2.2.2 Real-Time Decentralized Neural Backstepping Control Results

Trajectory tracking results are presented in Fig. 7.18. The tracking performance can
be verified for each of plant outputs, respectively. Figure 7.19 displays the angular
position tracking errors for joints 1–5.

The applied torques to each joint are shown in Fig. 7.20. All control signals are
always inside the prescribed limits given by the actuators manufacturer, that is, their

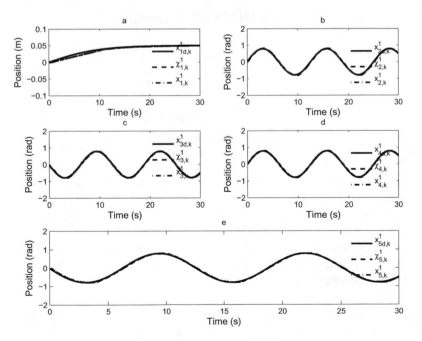

Fig. 7.15 Identification and tracking for each joint using DNBC algorithm

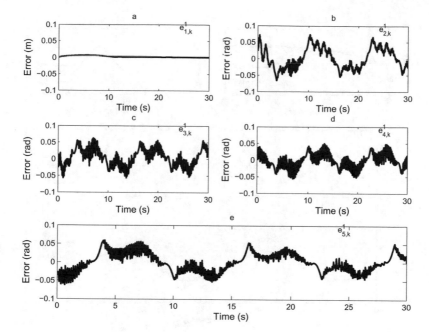

Fig. 7.16 Tracking errors for joints 1 5 (DNBC)

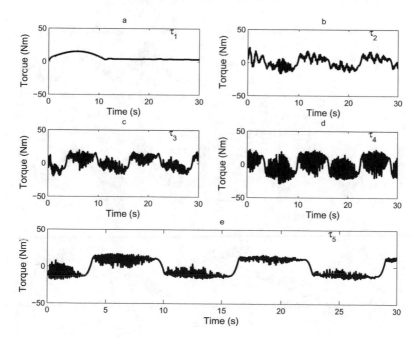

Fig. 7.17 Applied torques for joints 1–5 (DNBC)

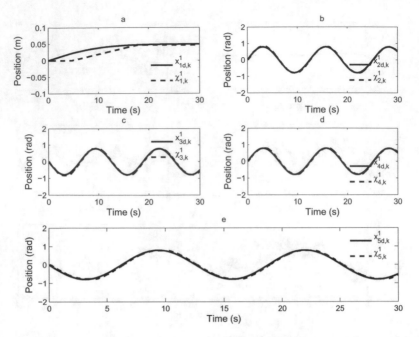

Fig. 7.18 Trajectory tracking for each joint using DNBS algorithm

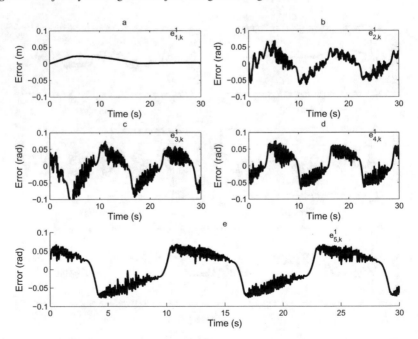

Fig. 7.19 Tracking errors for joints 1–5 (DNBS)

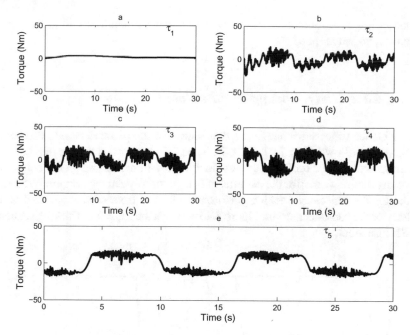

Fig. 7.20 Applied torques for joints 1–5 (DNBS)

Table 7.6 Comparison of the MSE for the real joint positions 5-DOF robot

Algorithm	$e_{1,k}^1$	$e_{2,k}^1$	$e_{3,k}^1$	$e_{4,k}^1$	$e_{5,k}^1$
DNBC	1.1127e-5	1.0332e-3	6.8420e-4	4.8675e-4	6.4256e-4
DNBS	1.4905e-4	1.8864e-3	1.8407e-3	1.2894e-3	1.8743e-3

absolute values are smaller than the bounds τ_1^{max} and τ_5^{max}, respectively. It is important to remark that we do not use any type of filter for the torque signals; thus, these signals are applied directly to each DC motor.

A comparative analysis of the two proposed schemes is included in Table 7.6.

According to the mean square error presented above, the scheme with better performance is the one based on the block control and sliding mode technique; on the other hand the scheme with slightly lower performance is the one based on the backstepping technique. It is important to remark that both schemes present an adequate performance for the case of trajectory tracking. Both algorithms have a complexity of order $O(n^2)$, however computational requirements are lower for the backstepping technique due to the fact that the neural identifier within the control scheme is no longer need.

7.2.3 Case Study: Shrimp Mobile Robot

The reference trajectories are selected as shown in the pictures below. Results are displayed as follows: Figs. 7.21, 7.22, 7.23 and 7.24 present on the top the trajectory tracking performance for the angular velocities of i-th motor and on the bottom the trajectory tracking error for the currents of i-th motor respectively. Figure 7.25 shows on the top the trajectory tracking performance for the front steer motor and on the bottom for the back steer motor. Figure 7.26 portrays the trajectory tracking sequence for Shrimp robot.

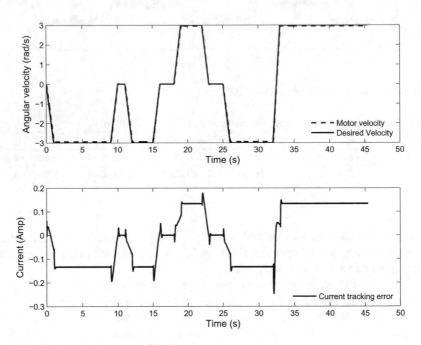

Fig. 7.21 Tracking performance of the front motor

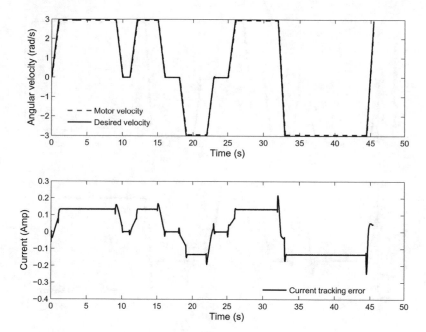

Fig. 7.22 Tracking performance of the right motors

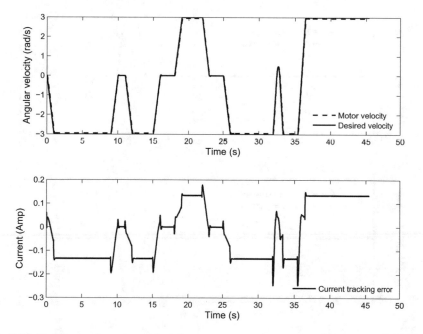

Fig. 7.23 Tracking performance of the left motors

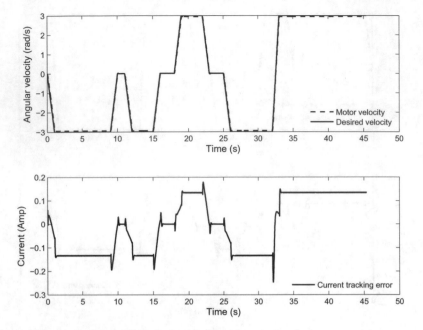

Fig. 7.24 Tracking performance of the back motor

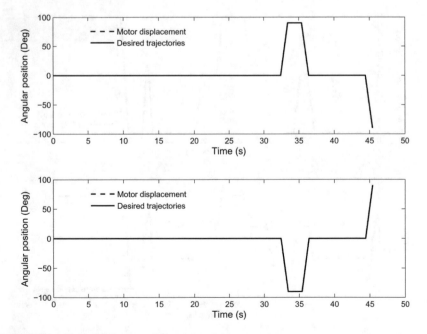

Fig. 7.25 Tracking performance of the steer motors

Fig. 7.26 Trajectory tracking sequence for Shrimp robot

7.3 Simulation Results

In this section simulation results for a seven DOF Mitsubishi PA10-7CE robot arm and KUKA youBot mobile robot are presented. The implemented control structures include the decentralized neural identification and control scheme presented in Chap. 3, the decentralized neural backstepping approach analyzed in Chap. 4 and the decentralized inverse optimal control for stabilization using a CLF approach proposed in Chap. 5.

7.3.1 Case Study: Seven DOF Mitsubishi PA10-7CE Robot Arm

For simulations, we select the following discrete-time trajectories [17]

Table 7.7 Parameters for desired trajectories

Link	c	d	ω (rad/s)
1	$\pi/2$	0.001	0.285
2	$\pi/3$	0.001	0.435
3	$\pi/2$	0.01	0.555
4	$\pi/3$	0.01	0.645
5	$\pi/2$	0.01	0.345
6	$\pi/3$	0.01	0.615
7	$\pi/2$	0.01	0.465

$$
\begin{aligned}
x^1_{1\mathrm{d},k} &= c_1(1 - e^{d_1 kT^3})\sin(\omega_1 kT)[\text{rad}] \\
x^1_{2\mathrm{d},k} &= c_2(1 - e^{d_2 kT^3})\sin(\omega_2 kT)[\text{rad}] \\
x^1_{3\mathrm{d},k} &= c_3(1 - e^{d_3 kT^3})\sin(\omega_3 kT)[\text{rad}] \\
x^1_{4\mathrm{d},k} &= c_4(1 - e^{d_4 kT^3})\sin(\omega_4 kT)[\text{rad}] \\
x^1_{5\mathrm{d},k} &= c_5(1 - e^{d_5 kT^3})\sin(\omega_5 kT)[\text{rad}] \\
x^1_{6\mathrm{d},k} &= c_6(1 - e^{d_6 kT^3})\sin(\omega_6 kT)[\text{rad}] \\
x^1_{7\mathrm{d},k} &= c_7(1 - e^{d_7 kT^3})\sin(\omega_7 kT)[\text{rad}]
\end{aligned}
\tag{7.18}
$$

the selected parameters c, d and ω for desired trajectories of each joint are shown in Table 7.7. The sampling time is selected as $T = 1$ ms.

These selected trajectories (7.18) incorporate a sinusoidal term to evaluate the performance in presence of relatively fast periodic signals, for which the non-linearities of the robot dynamics are really important

7.3.1.1 Decentralized Neural Block Control Results

Simulation results for identification and trajectory tracking using the decentralized neural block control (DNBC) scheme are shown in Figs. 7.27, 7.28, 7.29, 7.30, 7.31, 7.32 and 7.33. The initial conditions for the plant are different that those of the neural identifier and the desired trajectory.

The tracking error performance for each joint are presented in Fig. 7.34. Weights evolution for each joint is illustrated in Fig. 7.35. The applied torques for each joint are shown in Fig. 7.36. The control signals present oscillations at some time instants due to gains and fixed parameters selected for each controller. It is easy to see that all control signals are always inside of the prescribed limits given by the actuators manufacturer; that is, their absolute values are smaller than the bounds $\tau_1^{\max}$ to $\tau_7^{\max}$, respectively.

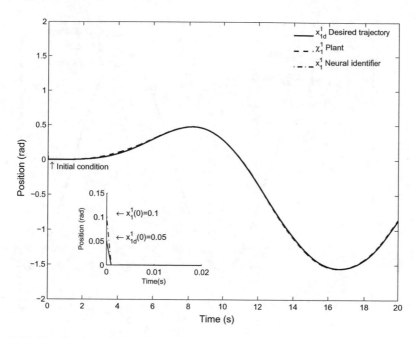

Fig. 7.27 Identification and tracking for joint 1

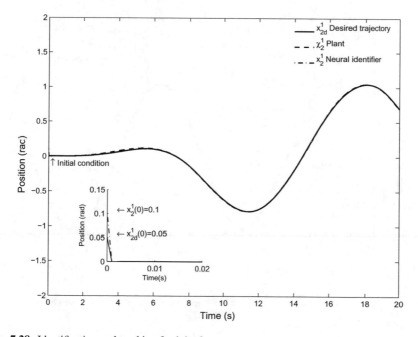

Fig. 7.28 Identification and tracking for joint 2

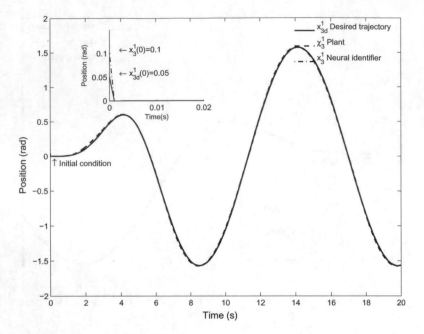

Fig. 7.29 Identification and tracking for joint 3

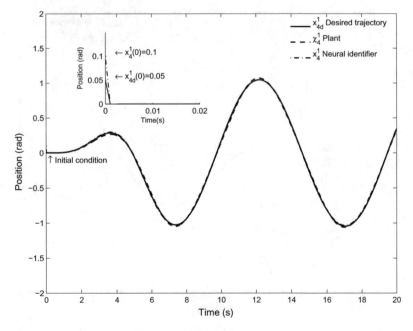

Fig. 7.30 Identification and tracking for joint 4

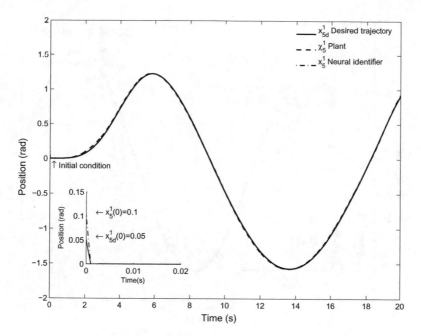

Fig. 7.31 Identification and tracking for joint 5

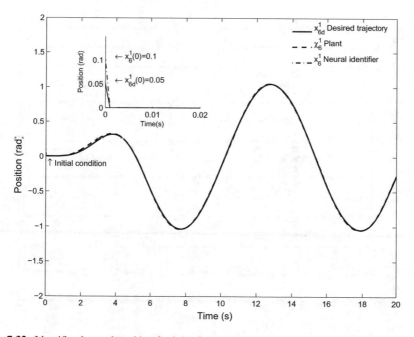

Fig. 7.32 Identification and tracking for joint 6

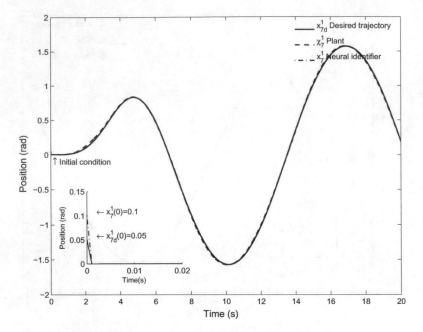

Fig. 7.33 Identification and tracking for joint 7

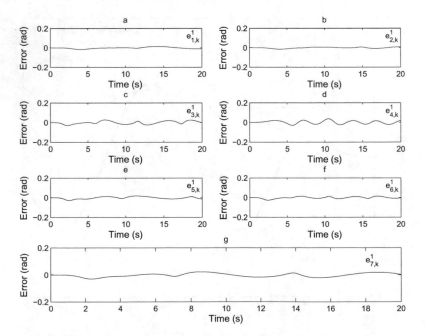

Fig. 7.34 Tracking errors for joints 1–7

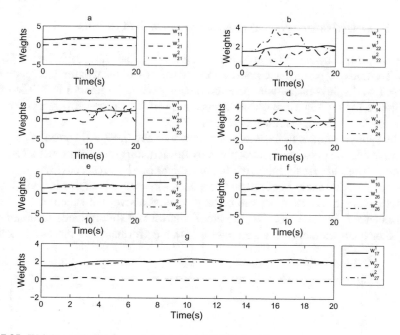

Fig. 7.35 Weights evolution for each joint using DNBC algorithm

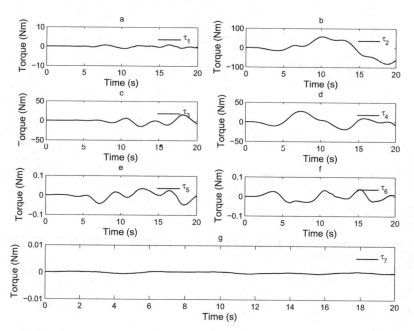

Fig. 7.36 Applied torques for joints 1–7

7.3.1.2 Decentralized Neural Backstepping Control Results

Trajectory tracking results are presented in Figs. 7.37, 7.38, 7.39, 7.40, 7.41, 7.42 and 7.43. The tracking performance can be verified for each of plant outputs, respectively. Figure 7.44 displays the angular position tracking errors for joints 1 to 7. The initial conditions for the plant are different that those of the desired trajectory. Weights evolution for each joint is illustrated in Fig. 7.45.

The applied torques to each joint are shown in Fig. 7.46. All control signals are always inside the prescribed limits given by the actuators manufacturer, that is, their absolute values are smaller than the bounds τ_1^{max} and τ_7^{max}, respectively.

A comparative analysis of the two proposed schemes is included in Table 7.8.

According to the mean square error presented above, the scheme with better performance is the one based on the block control and sliding mode technique; on the other hand the scheme with slightly lower performance is the one based on

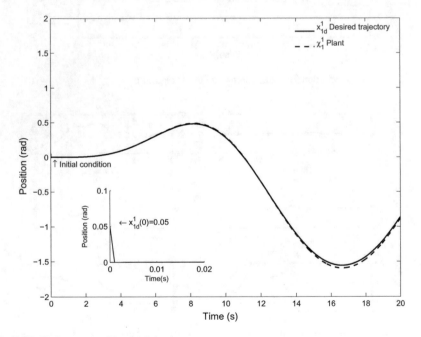

Fig. 7.37 Trajectory tracking for joint 1

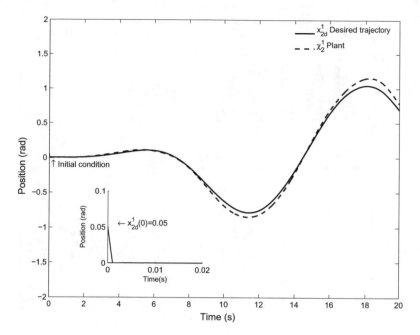

Fig. 7.38 Trajectory tracking for joint 2

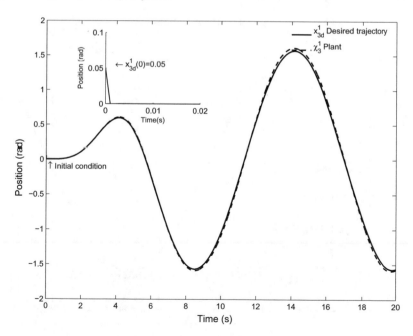

Fig. 7.39 Trajectory tracking for joint 3

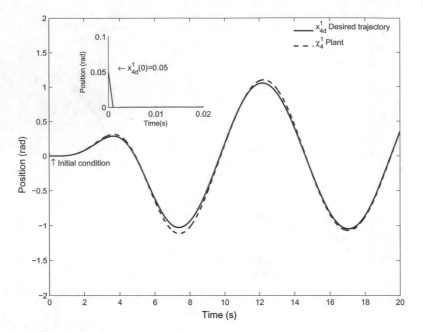

Fig. 7.40 Trajectory tracking for joint 4

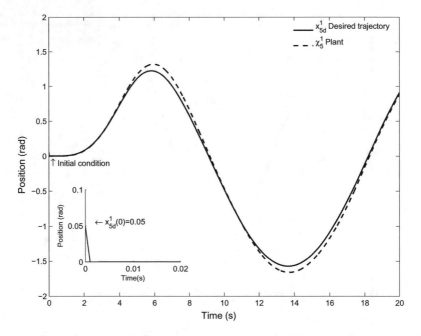

Fig. 7.41 Trajectory tracking for joint 5

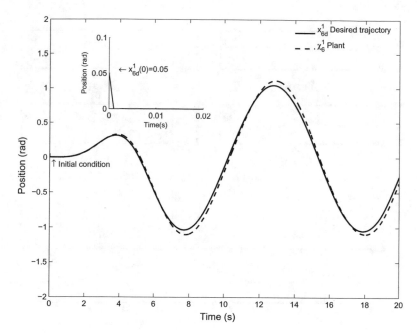

Fig. 7.42   Trajectory tracking for joint 6

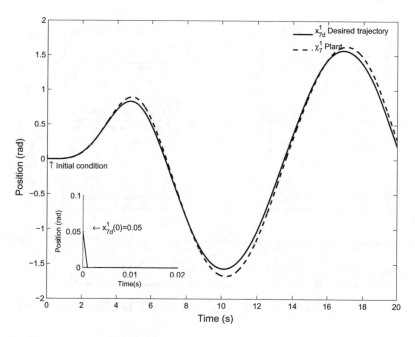

Fig. 7.43   Trajectory tracking for joint 7

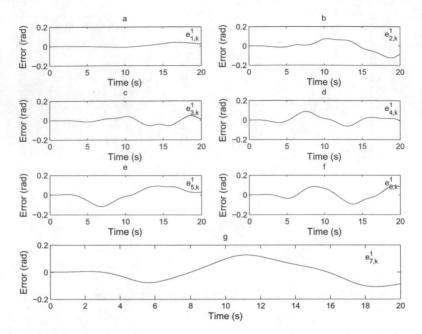

Fig. 7.44 Tracking errors for joints 1–7

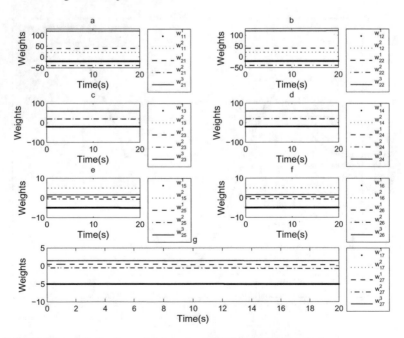

Fig. 7.45 Weights evolution for each joint using DNBS algorithm

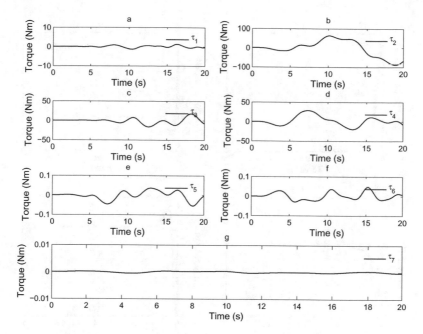

Fig. 7.46 Applied torques for joints 1–7

Table 7.8 Comparison of the MSE for the real joint positions 7-DOF robot

Algorithm	$e_{1,k}^1$	$e_{2,k}^1$	$e_{3,k}^1$	$e_{4,k}^1$	$e_{5,k}^1$	$e_{6,k}^1$	$e_{7,k}^1$
DNBC	6.5249e-5	4.1362e-5	2.2253e-4	2.4993e-4	1.3198e-4	9.4548e-5	1.8754e-4
DNBS	3.0900e-4	0.0029	8.3493e-4	0.0013	0.0038	0.0027	0.0047

the backstepping technique. It is important to remark that both schemes present an adequate performance for the case of trajectory tracking. Both algorithms have a complexity of order $O(n^2)$, however computational requirements are lower for the backstepping technique due to the fact that the neural identifier within the control scheme is no longer need.

In the literature, there are decentralized control algorithms in continuous-time [9, 18], however the order of complexity is higher due to the used numerical method. Furthermore, regarding the complexity issue, discrete-time results are available only for SISO systems [6, 19]. Other control schemes can not be compared because there are interconnections involved which are not rather quantifiable [5, 8, 12, 14].

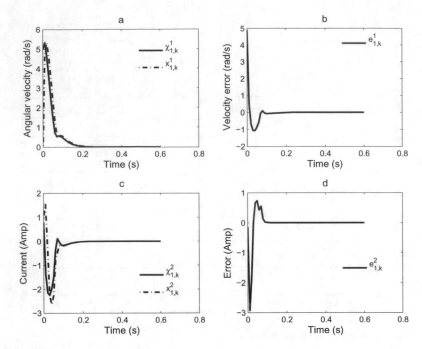

Fig. 7.47 Stabilization of the front right motor

7.3.2   Case Study: KUKA YouBot Mobile Robot

In this section, simulation results illustrate the effectiveness of the proposed method. Simulations are conducted using Matlab®[2] and Webots®[3] software. In this example, the system is composed of four subsystems. Every subsystem has the dynamics of a DC motor. The simulation consists on stopping the KUKA robot from initial conditions: angular velocity $\chi_{i,k}^1 = 5$ [rad/s] and armature current $\chi_{i,k}^2 = 0.8$ [Amp].

The results are displayed as follows: Figs. 7.47, 7.48, 7.49 and 7.50 present on (a) angular velocity stabilization for each i-th motor, (b) the velocity error, (c) current stabilization for each i-th motor, and (d) the respective error. The decentralized control inputs u_i are portrait in Fig. 7.51, where the (a)–(d) correspond to each one of them. From this results, it is clear that the decentralized control laws successfully stabilized the proposed robot at its equilibrium point.

---

[2]Matlab® is a registered trademark of MathWorks, http://www.mathworks.com/products/matlab/.
[3]Webots® is a registered trademark of Cyberbotics, https://www.cyberbotics.com/.

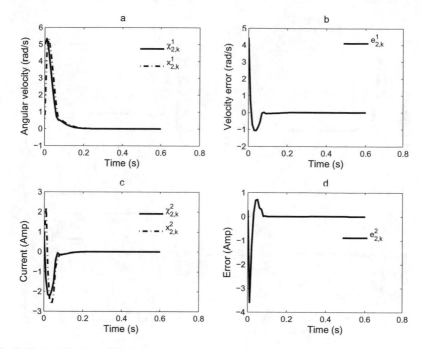

Fig. 7.48 Stabilization of the front left motor

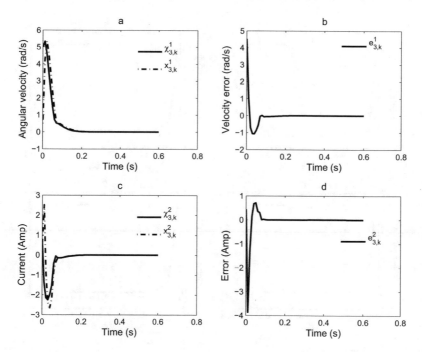

Fig. 7.49 Stabilization of the back right motor

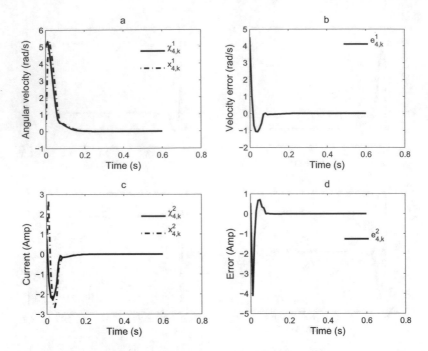

Fig. 7.50 Stabilization of the back left motor

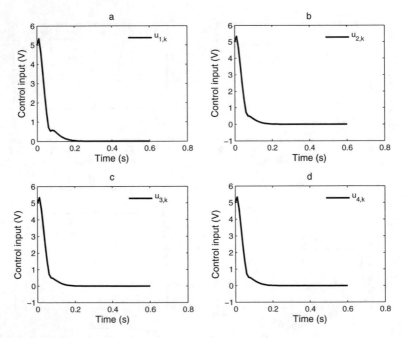

Fig. 7.51 The input voltage u_i

References

1. Ch, U.M, Babu, Y.S.K., Amaresh, K.: Sliding Mode Speed Control of a DC Motor. In: Proceedings of the International Conference on Communication Systems and Network Technologies (CSNT), pp. 387–391. Katra, Jammu, India (2011)
2. Feldkamp, L.A., Prokhorov, D.V., Feldkamp, T.M.: Simple and conditioned adaptive behavior from Kalman filter trained recurrent networks. Neural Netw. 16(5), 683–689 (2003)
3. Haykin, S.: Kalman Filtering and Neural Networks. Wiley, New York (2001)
4. Higuchi, M., Kawamura, T., Kaikogi, T., Murata, T., Kawaguchi, M.: Mitsubishi Clean Room Robot. Tecnical Review, Mitsubishi Heavy Industries Ltd (2003)
5. Huang, S., Tan, K.K., Lee, T.H.: Decentralized control design for large-scale systems with strong interconnections using neural networks. IEEE Trans. Autom. Control 48(5), 805–810 (2003)
6. Jagannathan, S.: Decentralized discrete-time neural network controller for a class of nonlinear systems with unknown interconnections. In: Proceedings of the IEEE International Symposium on Intelligent Control, pp. 268–273. Cyprus, Limassol (2005)
7. Jamisola, R.S., Maciejewski, A.A., Roberts, R.G.: Failure-tolerant path planning for the PA-10 robot operating amongst obstacles. In: Proceedings of IEEE International Conference on Robotics and Automation, pp. 4995–5000. New Orleans, LA, USA (2004)
8. Jin, Y.: Decentralized adaptive fuzzy control of robot manipulators. IEEE Trans. Syst. Man Cybern. Part B 28(1), 47–57 (1998)
9. Karakasoglu, A., Sudharsanan, S.I., Sundareshan, M.K.: Identification and decentralized adaptive control using dynamical neural networks with application to robotic manipulators. IEEE Trans. Neural Netw. 4(6), 919–930 (1993)
10. Kelly, R., Santibañez, V.: Control de Movimiento de Robots Manipuladores. Pearson Prentice Hall, Madrid, Espaa (2003)
11. Kennedy, C.W., Desai, J.P.: Force feedback using vision. In: Proceedings of the 11th International Conference on Advanced Robotics. Coimbra, Portugal (2003)
12. Liu, M.: Decentralized control of robot manipulators: nonlinear and adaptive approaches. IEEE Trans. Autom. Control 44(2), 357–363 (1999)
13. Mitsubishi Heavy Industries, Ltd, Instruction manual for installation, maintenance & safety. General Purpose Robot PA10 series, document SKG-GC20004 (2002)
14. Ni, M.-L., Er, M.J.: Decentralized control of robot manipulators with coupling and uncertainties. In: Proceedings of the American Control Conference, pp. 3326–3330. Chicago, IL, USA (2000)
15. Oonishi, K.: The open manipulator system of the MHIPA-10 robot. In: Proceedings of 30th International Symposium on Robotics. Tokio, Japan (1999)
16. Pholsiri, C.: Task Decision Making and Control of Robotic Manipulators. Ph.D. thesis, The University of Texas at Austin, Austin, TX, USA (2004)
17. Ramirez, C.: Dynamic modeling and torque-mode control of the Mitsubishi PA10-7CE robot. Master Dissertation (in Spanish), Instituto Tecnolgico de la Laguna, Torreon, Coahuila, Mexico (2008)
18. Sanchez, E.N., Gaytan, A., Saad, M.: Decentralized neural identification and control for robotics manipulators. In: Proceedings of the IEEE International Symposium on Intelligent Control, pp. 1614–1619. Germany, Munich (2006)
19. Spooner, J.T., Passino, K.M.: Decentralized adaptive control of nonlinear systems using radial basis neural networks. IEEE Trans. Autom. Control 44(11), 2050–2057 (1999)

Chapter 8
Conclusions

8.1 Conclusions

In this book, four decentralized control schemes are described as follows:

- The first *indirect* decentralized control scheme is designed with a modified recurrent high order neural network, which is able to identify the robot dynamics. The training of each neural network is performed on-line using an extended Kalman filter in a series-parallel configuration. Based on this neural identifier and applying the discrete-time block control approach, a nonlinear sliding manifold is formulated.
- The second *direct* decentralized neural control scheme is based on the backstepping technique, approximated by a high order neural network. The training of the neural network is performed on-line using an extended Kalman filter.
- The third decentralized control scheme is designed with a modified recurrent high order neural network, which is able to identify the robot dynamics. The training of each neural network is performed on-line using an extended Kalman filter in a series-parallel configuration. Based on this neural identifier a decentralized neural inverse optimal control for stabilization is applied. A simulation example illustrates the applicability of the proposed control technique.
- The fourth decentralized neural inverse optimal control is designed for trajectory tracking. The training of the neural network is performed on-line using an extended Kalman filter.
- All control schemes only require knowledge of the robot model structure; however the plant state vector must be available for measurement. Real-time results validate the efficiency of the proposed schemes for trajectory tracking when applied to a two DOF robot manipulator, a five DOF redundant robot and to a Shrimp mobile robot. Simulations using a seven DOF Mitsubishi PA10-7CE robot arm and a KUKA youbot mobile robot show the effectiveness of the proposed control schemes.

© Springer International Publishing Switzerland 2017
R. Garcia-Hernandez et al., *Decentralized Neural Control: Application to Robotics*,
Studies in Systems, Decision and Control 96, DOI 10.1007/978-3-319-53312-4_8

Printed in the United States
By Bookmasters